INTERNATIONAL MINING FORUM 2008

BALKEMA – Proceedings and Monographs
in Engineering, Water and Earth Sciences

International Mining Forum 2008
Economic Evaluation and Risk Analysis of Mineral Projects

Eugeniusz J. Sobczyk
Polish Academy of Sciences, Mineral and Energy Economy Research Institute, Cracow, Poland

Jerzy Kicki
AGH – University of Science and Technology, Department of Underground Mining, Cracow, Poland
Polish Academy of Sciences, Mineral and Energy Economy Research Institute, Cracow, Poland

Taylor & Francis
Taylor & Francis Group

LONDON / LEIDEN / NEW YORK / PHILADELPHIA / SINGAPORE

Published by: Taylor & Francis / Balkema
                  P.O. Box 447, 2300 AK Leiden, The Netherlands
                  e-mail: Pub.NL@tandf.co.uk
                  www.balkema.nl, www.tandf.co.uk, www.crcpress.com
ISBN10 0-415-46126-X
ISBN13 978-0-415-46126-9

Printed in the European Union

International Mining Forum 2008, Sobczyk & Kicki (eds) © 2008 Taylor & Francis Group, London, ISBN 978-0-415-46126-9

Table of Contents

International Mining Forum 2008, Sobczyk & Kicki (eds) © 2008 Taylor & Francis Group, London, ISBN 978-0-415-46126-9

Preface

The International Mining Forum was first held in 1998. The Forum is a meeting of scientists and professionals who, together with the organizers, establish ambitious aims to confront ideas and experiences, evaluate implemented solutions, and discuss new ideas that might change the image of the mining industry. The International Mining Forum is an event accompanying the School of Underground Mining, organized since 1992 under the auspices of AGH – University of Science and Technology and the Institute of Mineral Economy and Energy of the Polish Academy of Sciences of Cracow.

The motto of the School is the beautiful thought of professor Boleslaw Krupinski, an important and profound miner, who was the organizer and first Chairman of the International Organizing Committee of the World Mining Congress. He spoke as follows about the mining profession: "A miner always was, is and will be a man of technical and social progress. Only by improving the social and technical working conditions, a miner can discover the secrets and treasures of the Earth, conquer and exploit them for the benefit of all, reverse dangers caused by nature and provide the country with forces of nature".

Following these words we try to include papers addressing a broad range of subjects in the Forums' proceedings, in many occasions the problems are extending beyond underground mining.

For the fifth time a complete set of papers presented at the International Mining Forum will be published by the prestigious A.A. Balkema Publishers, owned by Taylor and Francis Group.

Also this time an interesting mix of subjects are treated in the proceedings, along with topics strictly connected with underground mining, e.g.:

- Economic Evaluation & Risk Analysis of Mineral Projects;
- New Trends in Coal Mine Methane (CMM) – Recovery and Utilization;
- Mining in the 21st Century – New Problems and Solutions.

Economic evaluation and risk analysis of mineral projects are always hot topics, inherent in mining activity. The session is devoted to many uncertainty and risk issues related to value and return and, therefore, affecting decision making processes.

The special session on recovery and utilization of coal mine methane focuses on upstream and downstream technical solutions for managing methane, best practices of effective management CMM emissions and financing methane drainage improvements and end-use options as well.

The organizers would like to express their gratitude to all persons, companies and institutions, who helped in bringing the Forum into being.

We hope that the Forum will contribute to the exchange of interesting experiences and establishing new acquaintances and friendships.

Jerzy Kicki
Chairman of the IMF 2008 Organizing Committee

International Mining Forum 2008, Sobczyk & Kicki (eds) © 2008 Taylor & Francis Group, London, ISBN 978-0-415-46126-9

Organization

Organizing Committee:
Jerzy Kicki (Chairman)
Eugeniusz J. Sobczyk (Secretary General)
Artur Dyczko
Jacek Jarosz
Piotr Saługa
Krzysztof Stachurski
Agnieszka Stopkowicz
Michał Kopacz

Advisory Group:
Prof. Volodymyr I. Bondarenko (National Mining University, Ukraine)
Mr. Wojciech Bradecki (Poland)
Prof. Jan Butra (CUPRUM Ltd., Poland)
Dr. Alfonso R. Carvajal (Universidad de La Serena, Chile)
Prof. Piotr Czaja (AGH – University of Science and Technology, Poland)
Prof. Józef Dubiński (Central Mining Institute, Poland)
Prof. Jaroslav Dvořáček (Technical University VSB, Czech Republic)
Prof. Paweł Krzystolik (Experimental Mine Barbara, Poland)
Prof. Garry G. Litvinsky (Donbass State Technical University, Ukraine)
Prof. Eugeniusz Mokrzycki (Polish Academy of Sciences, MEERI, Poland)
Prof. Roman Ney (Polish Academy of Sciences, MEERI, Poland)
Prof. Jacek Paraszczak (University of Laval, Canada)
Prof. Janusz Roszkowski (AGH – University of Science and Technology, Poland)
Prof. Stanisław Speczik, (Polish Geological Institute, Poland)
Prof. Anton Sroka (Technische Universität Bergakademie Freiberg, Germany)
Prof. Mladen Stjepanovic (University of Belgrade, Yugoslavia)
Prof. Antoni Tajduś (AGH – University of Science and Technology, Poland)
Prof. Kot F. v. Unrug (University of Kentucky, USA)
Dr. Leszek Wojno (Australia)
Dr Yuan Shujie (Anhui University of Science and Technology. Huainan, Anhui, The People's Republic of China)

International Mining Forum 2008, Sobczyk & Kicki (eds) © 2008 Taylor & Francis Group, London, ISBN 978-0-415-46126-9

International Trends in Standards and Regulations
for Valuation of Mining Industry Properties and Projects

Trevor R. Ellis

Mineral Property Valuer, Ellis International Services, Inc., Denver, Colorado, USA
Chairman, Extractive Industries Expert Group,
International Valuation Standards Committee, London

ABSTRACT: Real property valuation standards have been applied under limited circumstances to market valuations of minerals and petroleum properties for decades. Generally, these have been applied because the valuation was for use in court. This paper reviews the brief evolution of mineral valuation standards, from the 1995 publication of the first edition of The VALMIN Code in Australia, to the current edition of the International Valuation Standards. It provides a global perspective on the need for this set international standards. Recommendations are provided on the implementation of valuation standards within the extractive industries by government agencies, financial regulators, and professional societies, while advising on avoidance of conflicts between international and national standards. It addresses the shortage of valuers who are certified for valuation of minerals and petroleum industry assets using standards based on the Generally Accepted Valuation Principles (GAVP). Recommendations are made for valuer education and to solve licensing problems for working across borders.

KEYWORDS: Mineral valuation standards, regulations, International Valuation Standards, mineral valuer licensing, valuer education, appraiser, appraisal, financial reporting

1. MAJOR DEVELOPMENTS OF MINERAL VALUATION STANDARDS[1]

The history of standards specifically designed for mineral property valuation is brief, beginning only in 1995. However, valuation standards designed for real property valuation have been applied under certain circumstances to market valuations of minerals and petroleum properties for a few decades[2]. Much of this application of real property valuation standards has been for litigation purposes, particularly in the USA. There, the real property standards (Standards 1 and 2) of the Uniform Stan-

---

[1] The terms *valuation, evaluation,* and *appraisal* are often confused. *Valuation* is the process of estimating value. *Valuation* can also mean the value estimated by the *valuation* process. *Evaluation* is the process or diagnosis of condition or viability. In common use, the term *appraisal* has a similar meaning to *evaluation.* However, in the USA., the term *appraisal* is also used for what internationally is a *valuation* assignment and a formal *valuation* report. Similarly, a professional *valuer* is called an *appraiser* in the USA.
[2] Real property encompasses all of the rights, interests, and benefits associated with the ownership of real estate. Real property includes the many divisions of rights constituting the ownership of real estate, including rights of ownership of the minerals estate, or to access the minerals estate of land. Real property interests are normally recorded in a formal document such as a deed or lease.

dards of Professional Appraisal Practice (USPAP), have been applied to market valuation of interests in the minerals estate since publication of the first edition in 1987 [1]. In fact, the development of the United States Federal Government's Uniform Appraisal Standards for Federal Land Acquisitions, first published in 1971, has been to a large extent a compilation and analysis of a century of U.S. Federal court decisions on the compensation that the US Federal Government would pay for the compulsory acquisition of land and associated real property rights [2]. Many of the cases cited in the document involve disputes over the amount of "just compensation" that should be paid for minerals and petroleum interests, for which the courts prefer to base their decisions on market value evidence.

Prior to 1995, any instructions on minerals or petroleum valuation in valuation standards were minimal. Since then, there have been three major developments of mineral valuation standards.

1.1. *The VALMIN Code, Australia, 1995*

The Australasian Institute of Mining and Metallurgy (AusIMM) published its first edition of The VALMIN Code in 1995 [3]. This code introduced important ethical principles, which are to be applied to technical assessment and valuation work in expert reports that are developed for public or stock exchange (financial reporting) use.

The ethical principles are:
– transparency,
– materiality,
– competence,
– independence.

The 2005 Edition of The VALMIN Code is the second revision. This 23-page document is now titled, Code for the Technical Assessment and Valuation of Mineral and Petroleum Assets and Securities for Independent Expert Reports [4]. As with the previous editions, the major emphasis of the document remains on technical assessment. Application of the code is mandatory for members of the AusIMM and Australian Institute of Geoscientists, when developing expert reports for stock exchange filing purposes under Australian Corporation Law.

1.2. *CIMVal Standards, Canada, 2003*

In 2003, The Canadian Institute of Mining, Metallurgy and Petroleum (CIM) published, Standards and Guidelines for Valuation of Mineral Properties (CIMVal), a 33-page document developed by the CIM Special Committee on Valuation of Mineral Properties (CIMVal Committee) [5]. This document is the first set of mineral valuation standards to incorporate the three valuation approaches of the Generally Accepted Valuation Principles (GAVP) for market valuation, these being the Cost Approach, Sales Comparison Approach, and Income Capitalisation Approach. In the CIMVal document, the approaches are called, Cost Approach, Market Approach, and Income Approach.

The instructions within CIMVal are only for market valuation of the real property portion of mineral holdings (the most common subject of formal mineral valuations). The standards were developed for use by the Canadian mining industry in general, and for adoption by Canadian securities regulators and Canadian stock exchanges. The TSX Venture Exchange requires that the standards be followed when preparing valuations and valuation reports on mineral properties.

1.3. *International Valuation Standards, 2005*

In January 2005, the International Valuation Standards Committee (IVSC) published the Seventh Edition of the International Valuation Standards (IVSs). This edition of the IVSs was the first edition to include a set of valuation standards specifically for valuation of mineral and petroleum industry assets. This set of standards, now in the Eighth Edition of the IVSs, is designed for worldwide use, and draws upon the full set of IVSs to provide comprehensive GAVP-based standards for market and non-market valuations of all asset types within these industries. The Eighth Edition, published in 2007, has 462 pages [6].

2. THE INTERNATIONAL VALUATION STANDARDS

The standards contained within the IVSs cover the valuation of all asset types, both inside and outside the minerals and petroleum industries.

The IVSs classify assets into four primary property types:
– Real Property – includes ownership and interests in the surface estate of land and buildings, the minerals estate of mineral deposits and mineral rights, and surface and underground water.
– Personal Property – includes ownership and interests in property items that are not real estate, including physical items such as machinery and equipment, working capital, and securities.
– Businesses – Incorporated and non-incorporated entities pursuing an economic activity, including mine and quarry operating entities.
– Financial Interests – derived from the legal division of interests in businesses, real property, and personal property. These intangible assets include options to buy, lease, or sell real property and personal property.

The IVSs provide instructions for valuations for most uses, whether they be market-based or non-market valuations. An especially important use for valuations for which instructions are provided is private sector (non-government) financial reporting (corporate accounting), particularly with the International Financial Reporting Standards (IFRSs). Instructions are also provided for public sector (government) accounting, particularly with the International Public Sector Accounting Standards (IPSAS).

The standards composing the IVSs have been developed by the IVSC under a policy to provide mainly principles-based rather than rules-based standards. This policy is similar to that employed by the International Accounting Standards Board (IASB) in the development of the IFRSs. The policy minimizes both loopholes that might allow abuse of standards and obsolescence of these standards with changes in professional practice, regulations, and the economy.

In the IVSs, the foundation principles for valuation practice, the GAVP, are applied across all professional sectors of valuation. This consistency minimizes conflicts of theory, principles, and modes of application between valuation sectors. The benefits of this policy are conspicuous when the work of valuers from various sectors must be combined.

3. THE INTERNATIONAL VALUATION STANDARDS COMMITTEE

The IVSC published the first set of IVSs in 1985, this being mainly for valuation of land and buildings. Until now, the standards have all been developed by volunteers. The IVSC is a Non-Government Organization (NGO) member of the United Nations, having been granted Roster status with the United Nations Economic and Social Council in 1985 for consultation on valuation matters. The IVSC maintains liaison with international agencies, including the World Bank, the Commission of the European Union, the Bank of International Settlements (BIS), the Organization for Economic Cooperation and Development (OECD), the International Monetary Fund (IMF), the World Trade Organization (WTO), and the Basel Banking Committee. Fifty nations participate in developing and maintaining its standards. Each edition of the IVSs is translated into a number of languages.

4. NEED FOR A GLOBAL EXTRACTIVE INDUSTRIES VALUATION STANDARD

The following are some of the reasons why a global valuation standard for the extractive industries was found to be needed:
– Minerals and petroleum companies work internationally.
– Minerals and petroleum valuers work internationally.
– Companies report internationally to securities markets, investors, lenders, and government agencies.
– Investors and lenders compare projects internationally.
– Governments compete when leasing out, privatizing, or joint venturing mineral properties.
– More than 150 countries need standards, and it is highly desirable that the standards be compatible; and
– To support the possible use of current value accounting by extractive industry companies as they convert to the IFRSs for financial reporting.

5. DEVELOPMENT OF THE EXTRACTIVE INDUSTRIES VALUATION STANDARD

The IVSC's Extractive Industries Expert Group was formed in early 2001 to compose an expert submission to the IASB on minerals and petroleum industry valuation issues, pertaining to the planned development by the IASB of its extractive activities financial reporting standard. In the submission, the Group proposed that it draft an IVS extractive industries standard within a timeframe that would support the IASB's proposed extractive activities IFRS.

Current members of the Expert Group are from Australia, Canada, South Africa, and the USA. Current and previous members have been the leaders of the mineral valuation standards development initiatives of those countries. The members have actively participated in conferences on mineral valuation in many countries. The author has represented the IVSC in minerals and petroleum standards meetings at the United Nations, Geneva.

The extractive industries standard is an IVS Guidance Note, which supplements the body of IVS standards. During the drafting, input was obtained from the IASB Extractive Activities Project Team. A variety of important input was also obtained during the public exposure of the draft document in 2004, most of which was incorporated into the Guidance Note published in January 2005.

6. IVSC UNDERGOING CHANGE

The demands on the IVSC for its services are rapidly increasing, including the need to develop detailed guidance for more assets types, and the need for the IVSC to be able to provide professional interpretations of standards in a timely manner in response to requests. These demands mainly derive from increasing needs for the IVSC to support the expanding global adoption of the IFRSs and the resultant current value reporting of assets and liabilities. In 2008, the name of the IVSC will change from International Valuation Standards Committee to International Valuation Standards Council, as the organization undergoes a major transformation . Paid staff will gradually replace some of the volunteer professionals when stable sources of substantial funding are obtained. A professional board will also begin to address the criterion for the development of international valuer education programs.

7. IVSC BEST PRACTICE TECHNICAL PAPER

An IVSC extractive industries technical paper is expected to be released as an exposure draft for public comment in early 2008, with publication expected later in the year. This technical paper has been in development for three years by the IVSC's Extractive Industries Expert Group. The document will provide comprehensive, best practice guidelines to supplement the valuation standards of the Extractive Industries Guidance Note. The document explains the application and implications of many of the IVS standards to minerals and petroleum valuations.

8. STANDARDS TRENDS INTERNATIONALLY

Globalization is driving the need for global uniformity of standards in many business areas. Global uniformity is necessary in measurement and classification systems for all industry sectors for efficient, competitive international markets.

There is a rapidly progressing global convergence of private sector financial reporting standards to the IFRSs, supplemented by local guidance. An example of this convergence is that in February 2006, China issued 39 new IFRS-based accounting standards for business enterprises. Globalized stock markets, banking, and other financial sectors demand uniform financial reporting across national boundaries. These sectors will also demand uniform valuations of assets and liabilities for those uniform financial reports and many other uses. The IVSs are becoming the main source of uniform standards for those valuations.

A global trend that is particularly relevant to the minerals industry is the reduction from approximately 200 minerals and petroleum reserve-resource classification systems worldwide, towards a harmonized SPE-UNFC-CRIRSCO combined classification for both minerals and petroleum. The Society of Petroleum Engineers (SPE), other petroleum industry bodies, and the mining industry's Committee for Mineral Reserves International Reporting Standards (CRIRSCO) are working together to converge their petroleum and minerals classifications, and working with the United Nations to harmonize their petroleum and minerals classification systems with the United Nations Framework Classification (UNFC) for minerals and petroleum. Another trend that also appears to be taking place is the merging of real estate valuer certification bodies nationally and internationally, due to the developing need for uniform valuer education and certification globally.

9. MINERALS VALUATION TRENDS INTERNATIONALLY

Globalization of commodities markets, stock markets, mining and petroleum companies, financial services, and of the work of professionals within the extractive industries, will drive the following trends:
- Uniform minerals and petroleum valuation standards worldwide. The uniform standards will be supplemented by local requirements to meet distinct needs of states and provide enforcement of a code of ethics.
- Uniform qualification and competency requirements for minerals and petroleum valuers worldwide.
- Certification or licensing of those valuers by self-regulating organizations with demonstrated enforcement of a strong code of ethics.
- Increasing demand for extractive industries valuers to work across borders.
  At the same time, the following existing trends will likely continue for at least a number of years:
- A growing shortage of competent extractive industries valuers.
- Continued proliferation of national/state/provincial certification/license requirements for engineers, geologists, and land valuers.
- Inconsistencies in these requirements, lack of reciprocity between states, and varying land and mineral laws continue to increase the difficulty for extractive industries valuers to work legally across borders – for example, within the USA, Canada, and Australia.

Our professional societies, industries, and governments need to work towards achievement of the following for the benefit of minerals and petroleum industry professionals in general:
- Minimization of barriers to trade in professional services across state and national borders; and
- Implementation of a system of international reciprocity for certification and licensing of quailfied/competent professionals.

10. EXTRACTIVE INDUSTRIES VALUER EDUCATION

Minerals and petroleum valuation courses that teach market valuation based on the GAVP are rare worldwide. A few such courses have been taught by valuation societies and a university for continuing education purposes. No university program yet exists designed to provide comprehensive training in market valuation of extractive industries assets based on the GAVP. No comprehensive textbook is available for use for such education.
- Comprehensive minerals and petroleum valuation education must be made available.
- Instructional materials based on IVSs must be written.
- Appropriately qualified instructors must be groomed to teach these materials.
- Continuing education courses and comprehensive training programs are needed.

More extensive recommendations on this and related topics have been made in prior papers by this author [7], [8].

10.1. *Extractive Industries Valuer Education – Alternatives*

Students and interested professionals can learn much from published professional papers on extractive industries valuation. Some are available at www.mineralsappraisers.org and www.minevaluation.com.

A number of minerals industry professionals seeking education in market valuation have found it beneficial to take courses in real estate (real property) or business valuation. These courses are offered by professional valuation societies and educational institutions. Taking such courses may also help one qualify for valuer licensing, which is required in some jurisdictions to legally practice valuation of minerals or petroleum property as real property.

11. RECOMMENDATIONS

The following are the author's concluding recommendations:
- Nation states and national institutes should not develop their own valuation standards.
- Valuation standards setters should instead adopt the complete set of IVSs by reference (allowing automatic updating), to avoid valuation sector and international conflicts of standards and inappropriate rules.
- They should supplement the IVSs as needed for local conditions and to provide enforcement of an appropriate code of ethics.
- Support the development of educational materials and courses for teaching market-based extractive industries valuation.
- Remove or lower certification/licensing and other barriers,that prevent or impair entry by quailfied and competent extractive industries valuers to work in their specialization across state and national borders.

12. DISCLAIMER

The author is responsible for the analysis, opinions, and forecasts expressed in this paper, not the IVSC.

REFERENCES

[1] Appraisal Standards Board 2006: Uniform Standards of Professional Appraisal Practice and Advisory Opinions. 2006 Edition. The Appraisal Foundation, Washington, DC, 249 p. (available at: www.appraisalfoundation.org).

[2] Interagency Land Acquisition Conference 2000: Uniform Appraisal Standards for Federal Land Acquisitions. Appraisal Institute, 129 p. (available at: www.usdoj.gov/enrd/land-ack/).

[3] VALMIN Committee 1995: Code and Guidelines for Assessment and Valuation of Mineral Assets and Mineral Securities for Independent Expert Reports. Australasian Institute of Mining and Metallurgy, Melbourne, Australia, 19 p.

[4] VALMIN Committee 2005: Code for the Technical Assessment and Valuation of Mineral and Petroleum Assets and Securities for Independent Expert Reports. Australasian Institute of Mining and Metallurgy, Melbourne, Australia, 23 p. (available at www.mica.org.au).

[5] CIMVal Committee 2003: Standards and Guidelines for Valuation of Mineral Properties. CIM Special Committee on Valuation of Mineral Properties, Canadian Institute of Mining, Metallurgy and Petroleum, Montreal, Canada, 33 p. (available at www.cim.org).

[6] International Valuation Standards Committee 2007: International Valuation Standards, 8th Ed., London, 462 p.

[7] Ellis T.R. 2005: The International Valuation and Financial Reporting Standards – Their Content and Effect on Us. Symposium 2005: Window to the World, (Rhoden H.N., Steininger R.C. and Vikre P.G. eds), Geological Society of Nevada, Reno, Nevada, pp. 1297–1302.

[8] Ellis T.R. 2006: Implementation of the International Valuation Standards. Mining Engineering, Vol. 58, No. 2, February 2006, pp. 62–64.

International Mining Forum 2008, Sobczyk & Kicki (eds) © 2008 Taylor & Francis Group, London, ISBN 978-0-415-46126-9

About a Need for Polish Code for Mineral Asset Valuation

Jerzy Kicki
AGH – University of Science & Technology, Polish Academy of Sciences, MEERI, Cracow, Poland

Piotr Saługa
Polish Academy of Sciences, MEERI, Cracow, Poland

ABSTRACT: Polish mining industry has not developed yet complete standards for mineral (and/or petroleum) asset valuation. The problem of mineral asset valuation arose in the early '90s, caused by political system transformation. Unfortunately, Polish law has not kept up with transformation pace. Therefore in Poland, any kind of mineral asset valuation procedure or valuator competency rules haven't existed yet. This paper delivers a proposal of such a valuation standard called "POLVAL – Mineral Asset Valuation Code".

1. INTRODUCTION

In the past decades in the international markets have occurred some spectacular scandals connected with incompetent or biased mineral asset valuation. Their arise was possible through lack of uniform rules determining both principles of valuation process and competence and responsibility of mineral asset valuator. The direct result of those frauds were drastic falls in stock prices and bankruptcies of some mining companies (e.g. the collapse of Bre-X Minerals Ltd.).

In order to avoid such scandals in the future the most important mining countries have developped standards for competent mineral asset valuation. The aim of those documents is assurance that mineral asset valuation:
1. will be carried out by qualified persons;
2. will be accurate comprehensive and reliable;
3. all the essential information on mineral assets will be announced.

The first such a standard was developed in February 1995 in Australia. It has been called "Code for the Technical Assessment and Valuation of Mineral and Petroleum Assets and Securities for Independent Expert Reports" commonly known as the VALMIN Code [1].

An original intention of the code authors was preparing a document that would control mineral asset valuation in order to secure both mineral companies' and investors' rights and interests in Australian market. It turned out that the Code was more universal than one could ever suppose – the document, collecting best rules and principles, has become very practical for any mineral asset valuator in his work.

Complying with VALMIN regulations is obligatory for mineral and petroleum assets valuators – members of Australasian Institute of Mining and Metallurgy (AusIMM). It is also recommended by several institutions such as Australian Stock Exchange, Australian Securities & Investment Commission, Australian Institute of Geoscientists, etc.

The VALMIN Code is the best known valuation standard today. It has received wide recognition among mineral asset valuators. The code rules have served as a basis for other mineral asset valuation standards.

Introducing an analogous document in Canada in 2003 was met with applause from Canadian mineral industry. The document, called "Standards and Guidelines for Valuation of Mineral Properties" (CIMVal Code), was developed by Special Committee of Canadian Institute of Mining, Metallurgy and Petroleum [2].

Other mineral asset valuation standards have been developed in the USA and Republic of South Africa but due to various reasons they have not been introduced yet.

At present specialists from Extractive Industries Task Force of International Valuation Standards Committee (IVSC) have been working on an international valuation code.

2. WHAT IS THE CONTENT OF MINERAL (AND/OR PETROLEUM) ASSET VALUATION CODES?

In several important mining countries mineral asset valuation rules have been incorporated into the legislative system. It encompasses standardization of valuation work and certifying of qualified valuators. The regulations place a great importance on valuators' ethics and quality of their work. These are guaranteed thorough independent professional organizations of mineral asset valuators. A disqualification a valuator from the organization can definitely break his career.

The standards give a number of important definitions (such as a "value", "valuation", "expert" etc.) enabling codes' rules to be clear-cut, transparent and understandable. They include a requirement of disclosing of all crucial information about a valuation object. It is important to say that codes include the most essential regulations.

In principle they distinguish two groups of rules:
1. standards,
2. guidelines.

Following standards is strictly demanded whilst following guidelines – not. Nevertheless, they require reasons of departing from them.

The standards define independent valuation conditions, valuator and expert qualifications, general instructions and rules of valuation, etc.; they characterize a valuation commission, the process of pricing and the Valuation Report requirements.

All these issues have been detailed in the guidelines part – the codes give here instructions on what valuation approach and method should be used depending on project development stage.

3. SPECIFICS OF MINERAL ASSET VALUATION

Pricing of mineral (and/or petroleum) assets is a specific field of valuation activities. There are different goals of valuations.

The price is usually determined for the following cases:
– exploration lease/concession bids,
– initial public offerings,
– asset buying/selling prices, asset disposals,
– merger and acquisition transactions,
– stock market transactions,
– expropriation,
– litigation settlements,
– financial collateral,
– insurance claims,
– assessment of bank security,

- taxation purposes,
- securities reporting,
- accounting purposes, book value adjustments etc.

The reliable valuation is a sophisticated, multi-staged task.

This is because of specific features of mining investments, mainly connected with the deposit it-self:
- rarity of occurring;
- unique location;
- depletion problem;
- uncertainty and iniquity of deposit's volume, structure, and geologic characteristics.

Other characteristics of mineral assets are:
- exceptionally long pre-production (investment) period;
- long production (mining) period;
- diversified production conditions;
- capital intensivity;
- inflexibility of production process;
- unpredictability of mineral prices, connected with their cyclical character.

When deposit development starts immediately after its recovery and exploration, it is typically assumed that first cash flows occur after 3–5 years.

This causes the risk- and time value consequences. Fisher' theory of interest [3], [4] says that the time of receiving first incomes is crucial for the risky investments – the future cash flows of the long-term investments have a little value and there is no guarantee that the investor will receive them in the distant future.

The long investment period is then the crucial risk factor: there is a high uncertainty that econo-mic situation and mine conditions will stay the same. Mining investors must take into consideration that the initial decisions are often irreversible and their managerial flexibility is limited.

All the above mentioned specifics of mineral investments are sources of extremely high uncertain-ty and risk. An additional difficulty is caused by the fact that mineral asset must be valued at differ-rent stages of development, which often involves different approaches and methods.

The mineral asset valuation is then extremely specialized undertaking. It requires vast knowledge and experience in many fields – mainly geology, mining and economy, so that it can only be carried out by people with the appropriate skills and understanding of the above principles.

4. POLISH CONDITIONS

A need for mineral asset valuation arose gradually, after political system change in Poland in 1989.
    This necessity is connected with:
- a rise of mineral asset market;
- royalties and concessions requirements.

The current "Geological and Mining Law" interpretation says that the state is the owner of mineral deposits that are not a part of the land. The treasury is then the owner of mineral deposits that can be mined underground only. In this case there is not a possibility to trade a deposit. Those deposits are at the treasury's disposal. The rights of the treasury are executed by concession departments.

Deposits that can be mined by open-pit belong to the owners of the land [5].

Such an interpretation of law causes a range of consequences:
- open-pit properties are parts of the land, so that their valuation is in competence of a land app-raiser;
- there is no regulation that determines who might value an underground project.

As a rule of thumb one can assume the qualifications of a valuator. Most land appraisers do not have appropriate education. Thus the valuations of some deposits are incompetent and the results obtained are often false.

Even though the current regulations force land appraisers to turn to mineral expert, it is not determined what kind of specialist he should be, and what kind of qualifications he must have. There are no clear, uniform rules that would regulate these tasks.

The authors of this paper take the position that introducing a set of rules in the form of a code that would regulate mineral asset valuation in Poland is an urgent necessity.

The reasons for that statement are:
– growth of mineral transactions on national market;
– requirements from financial sector;
– necessity of reliable information for the treasury needs;
– lack of uniform criteria regulating mineral asset valuation in Poland (including lack of precisely described valuator qualifications);
– lack of rules determining the valuation approach depending on the development stage of the deposit;
– dispersion of Polish mineral asset valuators;
– lack of certifying of Polish experts.

Both Polish mining industry and financial market need mineral asset valuators. So far Polish experts, dispersed among different firms and institutions, have not got an independent, self-governing organization, which would enable them to organize, train and supervise valuators. In 2006 Polish Association of Mineral Asset Valuation (PAMAV) was founded. One of the main aims of the Association has been introduction of the Polish Code for Mineral Asset Valuation (called POLVAL), which would attribute valuation of deposits to qualified experts – Mineral Asset Valuators (TZKs).

In 2007 PAMAV formed a Special Committee on Polish Code for Mineral Asset Valuation.

Project POLVAL was approved in 2007. Using best practices and the experience of foreign codes, it also takes Polish specificity into account, e.g. ethics chapter has been included in the code. Obeying code rules by TZKs is to be strictly controlled by PAMAV association.

5. THE CONTENT OF THE POLISH CODE

The authors of the POLVAL code have taken a general position that the code's text should be compatible with foreign codes.

As described earlier [6], the POLVAL includes 5 chapters:
1. (A) A need for Polish code and the beginnings.
2. (D) Basic definitions.
3. (S) Standards – obligatory rules.
4. (W) Guidelines – recommended rules for valuation.
5. (Z) Ethical principles.

Chapter (A) presents introduction to the code. It includes the background and the review of foreign codes; it also describes the problem of ownership of deposits in Poland and the process of code arising.

Chapter (D) presents the main mineral valuation definitions such us "value", "valuation", "mineral asset valuator (TZK)", "expert", "valuation approach", "valuation method", "valuation process", "professional organization", "mineral", "mineral deposit", "resources", "reserves" etc.

"Standards" give the following principles that must be followed in the valuation process:
– materiality,
– transparency,

- independence,
- competence,
- reasonableness.

The code requires all the essential information for the reliable valuation needed. The notion "materiality" refers to data or information which "contribute to the determination of the mineral asset value, such that the inclusion or omission of such data or information might result in the reader of a Valuation Report coming to a substantially different conclusion as to the value of the mineral asset. By "transparent" POLVAL code means that the data and information used in (or excluded from) the valuation of a mineral asset must be set out clearly in the Valuation Report, along with the rationale for the choices and conclusions of the TZK. POLVAL determines the qualifications and responsibility the TZK, and the conditions of conducting an independent, reliable valuation. It includes general instructions for commissioning, valuation process and methodology, content of a Valuation Report and presenting valuation results.

Describing a process of valuation, "Standards" do not suggest a valuation approach and a valuation method, ceding the decision to the TZK. POLVAL justifies it as follows: "In the valuation process should be used at least two approaches. When the TZK is convinced that only one approach can be used he must explain the reasons why other approaches were not used. A choice of an approach should accomplish the goal of the valuation and – if it is possible – be agreed with a commissioning entity. In any case a risk analysis must be done".

POLVAL code recognizes the Polish resource/reserve reporting and classification system as a base for mineral asset valuation, adjusting it to foreign standards. Therefore the greatest importance is attached to the volumes of reserves and resources. Here it is important to say, that Polish mining industry needs to develop a Polish standard for reporting exploration results – similar to the Australian JORC code [7].

At the end "Standards" place a mandatory content of the Valuation Report. All actual volumes of reserves/resources must be presented and characterized in the report. The report must give dates and results of recent valuations, sources of main uncertainties and risks, and results of mineral property inspection(s).

The important parts of the Valuation Report are announcements of the TZK that he or she:
- is independent;
- is a qualified person;
- meets all the "Standards" requirements;
- meets – or not – the "Guidelines" requirements.

Finally, "Standards" give an mandatory content of the Report.

The "Guidelines" encompass all the rules ensuring an effective and reliable mineral asset valuation. Following "Guidelines" is not mandatory for the TZK but it is recommended by the code.

Perceiving great complexity of the valuation process, suitability and adequacy of individual valuation approaches and methods depending on project development stage [8], the POLVAL code distinguishes the following kinds of mineral assets:
- type i – early exploration assets;
- type ii – exploration assets;
- type iii – development assets;
- type iv – operating/abandoned mines;
- type v – mines at the closure stage.

It recommends the suitable approaches and the method of pricing.

Chapter (E), at last, gives principles of valuator's ethics, obligatory for the TZKs – members of PAMAV association. "Ethical principles" pay attention to independency and competence, stressing moral, personal, cultural and social matters.

SUMMARY

A correct and reliable mineral asset valuation is an extremely specialized, multi-staged undertaking. It is carried out at different stages of project development. It requires competence and vast experience in many fields – mainly mining, geology and economy.

Some mining countries have introduced valuation standards that regulate the process of mineral and/or petroleum asset valuation. In Poland there are not such regulations – especially there are no rules determining qualifications of a valuator. So that mineral asset valuation is often done by incompetent persons and valuation results are frequently biased. This problem has not been regulated yet.

There is no law in Poland regulating vertical delineation of land property. Therefore law interpretation grants an owner of the property a title to underlying mineral deposit. Consequently, a valuation of mineral deposit mined by open-pit are in competence of real estate valuators. Unfortunately, they often lack suitable qualifications and experience in the field of mineral asset valuation.

Therefore one should meet the increasing needs of both financial and mining sectors and national treasury for ensuring reliability of mineral asset valuations. It is necessary to create a uniform mineral asset valuation code to be helpful to the valuators in their professional work.

Last year, newly founded Polish Association for Mineral Asset Valuation (PAMAV), approved the Polish Code for Mineral Asset Valuation, called POLVAL. It was created thanks to many people, interested in regulating mineral asset valuation. Developing this code its authors used texts of foreign codes – mainly Australian VALMIN and Canadian CIMVal that have been introduced successfully. The Polish code, respecting Polish specifics, is the synthesis of both codes mentioned. One can hope that it will have a chance to become a binding mineral asset valuation standard in Poland.

REFERENCES

[1] AusIMM (Australasian Institute of Mining and Metallurgy) 1995: Code and Guidelines for Technical Assessment and/or Valuation of Mineral and Petroleum Assets and Mineral and Petroleum Securities for Independent Expert Reports (VALMIN Code). 2005 Edition. http://www.aig.org.au/files/valmin_122005.pdf

[2] CIMVAL (Special Committee of the Canadian Institute of Mining, Metallurgy and Petroleum on Valuation of Mineral Properties 2003: Standards and Guidelines for Valuation of Mineral Properties. Final version. CIMVal Workshop, Montreal 2003. http://www.cim.org/committees/CIMVal_Final_Standards.pdf

[3] Fisher I. 1907: The Rate of Interest. New York, Macmillan.

[4] Fisher I. 1930: The Theory of Interest. New York, Macmillan.

[5] Uberman R., Uberman R. 2004: Podstawy prawne dla wyceny wartości złóż kopalin eksploatowanych metodą odkrywkową (in Polish). Mineral Resources Management, Special Issue, No. 1, wyd. IGSMiE PAN 2004.

[6] Saluga P. 2007b: A Proposal of Polish Code for Mineral Asset Valuation (in Polish). Materiały XVII Konferencji z cyklu „Aktualia i perspektywy gospodarki surowcami mineralnymi", wyd. IGSMiE PAN, Kraków 2007.

[7] Kicki J., Niec M. 2006: Na drodze do ujednolicenia klasyfikacji zasobów złóż w skali międzynarodowej (in Polish). Mineral Resources Management, Special Issue, No. 2, wyd. IGSMiE PAN 2006.

[8] Saluga P. 2007a: Mineral Asset Valuation – a Basic Tool for Effective Decision Taking in the Mining Industry (in Polish). Proceedings of the Polish Mining Congress 2007, Session 11, Economics & Management, Mineral Resources Management, Vol. 23, Special Issue, No. 2, wyd. IGSMiE PAN, Krakow 2007.

International Mining Forum 2008, Sobczyk & Kicki (eds) © 2008 Taylor & Francis Group, London, ISBN 978-0-415-46126-9

Development of the International Valuation Standards Committee's Best Practice Guidelines for Valuations in the Extractive Industries

Trevor R. Ellis
Mineral Property Valuer, Ellis International Services, Inc., Denver, Colorado, USA
Chairman, Extractive Industries Expert Group,
International Valuation Standards Committee, London

ABSTRACT: In early 2008, the International Valuation Standards Committee (IVSC) will release the exposure draft of its Extractive Industries Technical Paper for public comment. The document has been in development for almost four years. Development began many months before the associated standards in the Extractive Industries Guidance Note received final approval for 2005 publication. The primary purpose of the document is to provide best practice valuation guidelines for extractive industries applications of the International Valuation Standards (IVSs). The document interprets the meaning and implications of a number of important parts of the IVSs for their application to valuations within the minerals and petroleum industries. The document describes in detail the application of the Generally Accepted Valuation Principles to extractive industries valuation. It also provides best practice guidance on many technical matters for minerals and petroleum valuations. The Technical Paper is expected to be finalized and published during 2008 by the IVSC.

KEYWORDS: International Valuation Standards Committee, Extractive Industries Technical Paper, minerals and petroleum valuation standards, best practice

1. STATUS OF THE GUIDELINES DEVELOPMENT, DECEMBER 2007

As of the close of 2007, the Extractive Industries Expert Group of the International Valuation Standards Committee (IVSC) has been developing its best practice valuation guidelines document for the extractive industries of minerals and petroleum for almost 4 years. During 2007, final drafts of the document were reviewed twice by meetings of the full Standards Board of the IVSC. At the November 2007 meeting of the Standards Board in London, the document was approved for public exposure subject to further technical editing. The exposure draft of the document is expected to be posted for public exposure in early 2008 by the IVSC to seek public comment. The document is expected to be finalized and published as a Technical Paper by the IVSC during 2008. At the close of 2007, the 13,000 word draft Technical Paper carries the title, Valuation of Properties in the Extractive Industries.

2. HISTORY OF GUIDELINES DEVELOPMENT

Conceptual development of the Technical Paper began in March 2004, almost a year before the Extractive Industries Guidance Note of standards developed by the Expert Group was published in the Seventh Edition of the International Valuation Standards (IVSs) in February 2005. The Expert Group had begun drafting those standards in September 2002. The Guidance Note has been republished in the Eighth (2007) Edition of the IVSs as Guidance Note 14, also carrying the title, Valuation of Properties in the Extractive Industries [1].

The members of the all-volunteer Extractive Industries Expert Group remain unchanged from those at the time of the conceptual development of Technical Paper in March of 2004:
- Trevor Ellis, USA, Chairman,
- William Roscoe, Canada,
- Alastair Macfarlane, South Africa,
- Donald Warnken, USA,
- Raymond Westwood, Australia.

The initial plan submitted to the IVSC called for the Technical Paper to be developed on a fast-track schedule similar to that undertaken for the development of the standards of the Extractive Industries Guidance Note (GN), with publication of the paper planned for the second half of 2005. This was so that the paper would be available to the Extractive Activities Project Team of the International Accounting Standards Board (IASB) on a timely basis to reference. The Project Team is presently conducting research to develop recommendations for the IASB regarding the content of the soon to be drafted Extractive Activities International Financial Reporting Standard (IFRS) [2], [3].

Many factors conspired to slow the pace of drafting the Technical Paper. The research schedule of the IASB's Project Team has been extended by a number of years, removing most of the incentive for fast-track development of the Technical Paper. The complexities involved in writing a paper to meet a number of goals and changing expectations were more difficult than expected, even more difficult than drafting the GN. Also, members of the Expert Group (previously called the Extractive Industries Task Force) found it difficult to keep putting volunteer work for the Expert Group ahead of plentiful requests to do lucrative extractive industries valuation work, especially after having done that for drafting the GN and the 2001 submission to the IASB.

3. THE ROLE TO BE FULFILLED BY THE TECHNICAL PAPER

When planning the content for the Technical Paper, the Expert Group selected the following primary purposes for the document to fulfill:
- Provision of best practice guidelines to supplement the standards provided by the Extractive Industries GN in the IVSs.
- To teach the fundamental principles of minerals and petroleum industry valuation practice, including the GAVP, to interested extractive industry practitioners and users of extractive industries valuations, who have little or no prior knowledge of valuation principles. This purpose was added due to there being no textbook yet published that teaches these principles specifically for the extractive industries. The expert group expects that a large majority of readers of the Technical Paper will have little knowledge of these valuation principles.
- To bring to the reader's attention appropriate conduct and ethical principles that should be followed to aid the production of unbiased valuation reports that clearly convey all relevant information.
- To provide information about available certification of valuers for extractive industries valuation, and to provide guidance regarding appropriate education and experience for becoming qualified as a minerals or petroleum industry valuer, including possible sources for that education. This purpose was viewed as important due to the severe shortage globally of qualified, competent extractive industries valuers, and education and certification opportunities for persons wishing to become qualified.

– To advise the user of the Technical Paper about minerals and petroleum industry standards, practice, defined terms, and financial reporting requirements, that are widely used internationally and for which an understanding is fundamental to the development of and accurate valuation.

The expert group decided that the scope of the Technical Paper should remain limited in size so that it would maintain its character as a paper that provides best practice guidelines that supplement the standards of the IVSs. Therefore, although essential background information is provided throughout to aid the understanding of the guidelines, examples are rarely used. No illustrations, tables, example calculations, or data are included.

The drafting of the Technical Paper by the Expert Group through to its completion in March 2007, was directed to achieve all of these purposes within the limited size of the Technical Paper. The style and content of writing was designed for the wide readership envisaged. The completed draft document provided to the IVSC Standards Board for discussion at its April 2007 meeting in San Francisco was approximately 16,000 words in size and included an addendum devoted to minerals and petroleum valuer education and qualifications.

3.1. *Philosophical difficulties encountered*

Over the eight months since the completed draft of the Technical Paper was delivered to the Standards Board, the paper has undergone a number of rounds of editing by Standards Board members. The end result of the editing has been to focus the document entirely on the first of the above listed purposes: provision of best practice guidelines to supplement the standards provided by the Extractive Industries GN. Almost everything included for the other purposes that is not necessary for the first purpose has been deleted, resulting in a 20% reduction in the word count of the document.

Over the seven years since the author first convened the IVSC Extractive Industries Expert Group (Task Force) in February 2001, the IVSC has grown and matured considerably. This has paralleled the related growth and maturing of the IASB (effectively the IVSC's big sister body), with the global adoption of the IFRSs. Since author presented the draft Technical Paper to the Standards Board in April 2007 in San Francisco, the IVSC has begun transforming its management structure from that designed for a small, all-volunteer organization, into that of a well-funded organization, with full-time salaried management. At the same time, the recommendations of a critical review of the entire IVS standards book are being implemented, resulting in tightening of the structure and content of the book. This growth, maturation, and transformation of the IVSC, is reflected in the evolution of a recent strict opinion of the Standards Board for the limited purpose and content of the Technical Paper.

The Extractive Industries Technical Paper is only the second Technical Paper developed by the IVSC.

The other, Mass Appraisal for Property Taxation, was published in 2005. At the time of the conceptual planning of the Extractive Industries Technical Paper in 2004, the Expert Group was told by the IVSC that no structure had been defined for Technical Papers, and that the approval process would be "much less rigorous" than for the GN.

4. FINALIZATION OF THE TECHNICAL PAPER

Now that the purpose and content that is acceptable for the Technical Paper has been determined, the focus of interactions with the Standards Board members has moved towards to the finer aspects of theoretical and technical issues pertaining to extractive industries valuation standards and practice. Once these are settled, the exposure draft of the document should be released for public comment. Comments received will be reviewed by the Extractive Industries Expert Group for possible integration into the Technical Paper. Then, after another review by the Standards Board, the document should through the IVSC editor and management to publication. The form of publication will likely be left for determination in 2008 by either the interim IVSC management or the incoming management of the restructured IVSC.

REFERENCES

[1] International Valuation Standards Committee 2007: International Valuation Standards, 8th Ed., London, 462 p.
[2] Ellis T.R. 2005: The International Valuation and Financial Reporting Standards – Their Content and Effect on Us. Symposium 2005: Window to the World, (Rhoden H.N., Steininger R.C. and Vikre P.G. eds), Geological Society of Nevada, Reno, Nevada, pp. 1297–1302.
[3] Ellis T.R. 2006: Implementation of the International Valuation Standards. Mining Engineering, Vol. 58, No. 2, February 2006, pp. 62–64.

International Mining Forum 2008, Sobczyk & Kicki (eds) © 2008 Taylor & Francis Group, London, ISBN 978-0-415-46126-9

Standardized Measure of Discounted Future Net Cash Flows Related to Proved Oil and Gas Reserves and Capitalised Costs of Exploration and Development As Two Ways to Include Oil and Gas Assets' Valuation in Financial Statements

Robert Uberman
The Andrzej Frycz Modrzewski University College, Cracow, Poland
Executive Committee Member of the Polish Association of Mineral Assets Valuation

ABSTRACT: Financial Statements of the biggest four global oil companies are analyzed to indentify how Standardized Measure of Discounted Future Net Cash Flows Related to Proved Oil and Gas Reserves and Capitalized Costs of Exploration and Development influence their market value.

KEYWORDS: Mineral deposits, valuation methods, SMOG, oil companies' financial reporting

1. INTRODUCTION

It is needless to prove that mineral deposits (or rights for exploration and development) constitute the crucial part of mining companies' assets. However there is a certain difficulty to find explicite information in financial statements indicating their value, especially if an investor, quite logically, seeks it on introductory pages. On the contrary – searching such information might be as much difficult undertaking as exploring for mineral deposits themselves. Universal rules regarding this issue are still to be adopted, at least by these countries which hold substantial mineral deposits and simultaneously claim to have well developed financial markets[1]. These constitutes one of the most substantial exemption from IAS 16 Property, plant and equipment. A Standard dealing with issues exempted from IAS 38 Intangible Assets has not been published yet, either[2]. Simultaneously, the US GAAP does not provide any complex and harmonised regulation for this matter.

As a rule the following information on mineral deposits valuation may be found:
– proved reserve quantities;
– capitalized costs relating to exploration and development;
– costs incurred for property acquisition, exploration, and development activities;
– results of operations for mining activities (if these are defined as separate segments);
– a standardized measure of discounted future net cash flows relating to proved oil and gas[3].

Capitalised costs and standardized measure of discounted future net cash flows relating to proved oil and gas are the only two to reflect directly, at least to some extend, value of the mineral assets held. Consequently they will be discussed below.

---

[1] In most cases a list of such countries encompasses: Australia, Canada, RSA, USA and the United Kingdom. See: KPMG: *Mining – A survey of Global Reporting Trends 2003*, KPMG International, 6 p.
[2] *Międzynarodowe Standardy Sprawozdawczości Finansowej 2007*. Stowarzyszenie Księgowych w Polsce. Warszawa 2007, str. 1–317.
[3] Further referred as SMOG.

2. METHODOLOGY APPLIED TO DETERMINE STANDARDIZED MEASURE OF DISCOUNTED FUTURE CASH FLOWS

FASB 69 prescribes an obligatory reporting of the so called standardized measure of discounted future net cash flows relating to proved oil and gas reserves – SMOG[4], based on ASR No. 253: Adoption of Requirements for Financial Accounting and Reporting Practices for Oil and Gas Producing Activities of the Securities Exchange Board of the USA. It is important to point out that introduction of the discussed regulation was somewhat controversial and several important members of regulating bodies dissented. It was admitted that the methodology applied to determine SMOG is too vague and simplified to represent a market value of the oil and gas reserves. It will be treated as a simplified indicator of such value.

A methodology prescribed by US GAAP requires Companies to adopt several key assumptions which are listed below[5]:
- standardized measure of discounted future net cash flows from proved oil and gas reserves is amount of the future net cash flows less the computed discount;
- a rate and time of reserves depletion is assumed by the reporting company;
- prices applied are year-end prices of oil and gas – future price changes shall be considered only to the extent provided by contractual arrangements in existence at year-end;
- costs should be computed by estimating the expenditures to be incurred in developing and producing the proved oil and gas reserves at the end of the year, based on year-end costs and assuming continuation of existing economic conditions;
- a fixed 10% discount rate is applied;
- the appropriate year-end statutory tax rates, with consideration of future tax rates already legislated, less the tax basis of the properties involved, should be applied.

Yet, there is no obstacle preventing from its application to other minerals like sulphur already mentioned in the FASB 69.

The practical application of SMOG valuation depends on stability of the macroeconomic conditions. The pivotal role here is played by the discount rate. Since it is fixed at nominal value reporting companies are deprived from using the most commonly used tool in reflecting changing volatility of key assumptions. This regulation originates from 1970s when 10% discount rate could, at least to some extend, correspond to the that time prevailing hurdle rate in oil companies[6]. Application of uniform discount rate may be justified for the purpose of financial reporting where ability to compare date across the industry and independence from subjective judgments of reporting companies form essential prerequisites. But even in this framework such rigid requirement is hard to justify. There are many cases in financial reporting where a discount rate is identified by companies, e.g. employees benefits. This case calls for re-examination and new guidelines.

Nevertheless almost all listed companies dealing with oil and gas report SMOG and this refers also to companies not present on the US financial markets (e.g. OMV).

Extending use of the SMOG for other minerals shall be deemed as desirable for the following reasons:
- SMOG methodology has been used for nearly 30 years now and, although controversial at the adoption stage, is clearly a helpful tool for the investors;
- for mining companies mineral deposits constitute key assets and traditional approach based on historical costs tested against possible loss of value fails to reflect its real impact on mining companies valuation;
- disclosure of physical reserves only does not provide any indication of its value for companies.

---

[4] Standard Measure of Oil and Gas.

[5] FASB 69, pp. 30–34.

[6] For calculations made in real prices.

3. CAPITALISED EXPLORATION AND DEVELOPMENT COSTS

Capitalised exploration and development costs reflect historical cost approach to asset valuation, still the accepted by accounting standards and commonly used.

According to the KPMG study exploration and evaluation expenditures are those incurred in connection with acquisition of rights to explore, investigate, examine and evaluate an area for mineralization.

Exploration may be conducted before or after the acquisition of mineral rights[7]. IFRS 6 mentions the following examples of expenditures (the list is not exhaustive):
- acquisition of rights to explore;
- topographical, geological, geochemical and geophysical studies;
- exploratory drilling;
- trenching;
- sampling; and
- activities in relation to evaluating the technical feasibility and commercial viability of extracting a mineral resource[8].

An important restriction has been added forbidding application of IFRS 6 to expenditures incurred:
- before the exploration for and evaluation of mineral resources, such as expenditures incurred before the entity has obtained the legal rights to explore a specific area;
- after the technical feasibility and commercial viability of extracting a mineral resource are demonstrable[9].

According to the KPMG study development costs typically include those incurred for the design of the mine plan, obtaining the necessary permits, constructing and commissioning the facilities and preparation of the mine and necessary infrastructure for production. The mine development phase generally begins after completion of a feasibility study and ends upon the commencement of commercial production.

The last provision, although makes impression of being precise, in practice leaves space for various interpretations. Some companies adopt approach in which commencement of commercial production is defined not in terms of real beginning of operations but in terms of ability to do so. Some other try to set a cut off point at the beginning of underground works.

KPMG identified the following practices while investigating the top 44 (50) world mining companies:

Table 3.1. Accounting treatment of exploration and evaluation expenditure in 2006 (2003 data in brackets)[10]

	Australia	Canada	RSA	UK	USA	Other	Total
Expenses as incurred	1 (0)	0 (1)	0 (1)	1 (0)	6 (3)	2 (2)	10 (7)
Expensed as incurred until the ore body is deemed commercially recoverable, at which time all subsequent costs are deferred	3 (2)	12 (11)	6 (5)	4 (4)	0 (2)	2 (1)	27 (25)
General exploration costs are expensed as incurred and exploration costs on specific projects are deferred	0 (0)	0 (0)	0 (1)	0 (0)	0 (0)	0 (1)	0 (2)

[7]KPMG: *Global Mining Reporting Trends 2006*. KPMG International, 15 p.
[8]IFRS 6, 9 p.
[9]IFRS 6, 5 p.
[10]KPMG: *Global Mining Reporting Trends 2006*. KPMG International, 15 p. and KPMG; *Mining – A Survey of Global Reporting Trends 2003*. KPMG International, 38 p.

Table 3.1. Cont'd

	Australia	Canada	RSA	UK	USA	Other	Total
Capitalise until a reasonable assessment can be made of the existence of reserves	1 (3)	0 (0)	0 (0)	2 (1)	0 (0)	2 (0)	5 (4)
Capitalised	0 (1)	0 (0)	0 (0)	0 (0)	0 (0)	0 (0)	0 (1)
Police not disclosed	0 (0)	0 (1)	0 (0)	0 (0)	1 (2)	1 (8)	2 (11)
Total	5 (6)	12 (13)	6 (7)	7 (5)	7 (7)	7 (12)	44 (50)

Despite differences it remains clear that the approach based on IFRS 6 prevails among top mining companies – it had been adopted by nearly 2/3 of these who disclosed their policy.

Development costs, similarly to the exploration ones are treated differently by the top 50. Differences identified by KPMG are presented below.

Table 3.2. Accounting treatment of development costs[11]

	Australia	Canada	RSA	United Kingdom	USA	Other	Total
Capitalised	3 (4)	0 (5)	1 (3)	4 (5)	3 (1)	3 (4)	14 (22)
Development costs incurred to maintain current production are expensed, while development ore bodies and development in advance of production are capitalized	1 (2)	12 (7)	4 (3)	2 (0)	4 (5)	3 (2)	26 (19)
Policy not disclosed	1 (0)	0 (1)	1 (1)	1 (0)	0 (1)	1 (6)	4 (9)
Total	5 (6)	12 (13)	6 (7)	7 (5)	7 (7)	7 (12)	44 (50)

Capitalised exploration and development costs can hardly be used as an indicator of mineral assets valuation since they carry all conceptual limitations of the cost approach in general. Since they are tested against market value only for purpose of impairment recognition investors can reasonably expect them to reflect only a minimal value mineral assets. But even such interpretation may lead to false conclusions. "Minimal" in this sense does not mean a value under the least favourable circumstances but rather a random combination of asset valuation dependent on how and when they have been acquired.

4. REVIEW OF SMOG AND CAPITAL COSTS DISCLOSERS ON MARKET VALUATION OF THE SELECTED OIL AND GAS COMPANIES

The most important oil and gas companies of the world have been examined with regard to disclosures made in their respective financial statements on SMOG, exploration and development costs capitalisation and possible impact on their market capitalisation.

---

[11] KPMG: *Global Mining Reporting Trends 2006*. KPMG International, 19 p. and KPMG; *Mining – A Survey of Global Reporting Trends 2003*. KPMG International, 40 p.

British Petroleum, ChevronTexaco, ExxonMobil and Royal Dutch Shell were selected for this purpose as the ones to combine following features:

– they are the four biggest companies of the industry in terms of capitalisation and revenues[12];
– they all originate from the leading mining countries;
– they have been listed on NYSE at time when IFRS 69[13] (providing for use of SMOG) was introduced – therefore both companies and investors have had quite a sufficient time to accommodate to the SMOG disclosure.

SMOG capitalised costs of exploration and development activities and oil and gas reserves in natural measures were compared to market value (MV) and book value (BV) in order to identify their impact on investors' valuation of the companies in consideration.

The review covers years 2000–2006. Data were derived from financial statements of the respective companies presented mostly in their annual reports[14]. Data from the most recent financial statement were used regarding all years covered (e.g. 2004 Chevron data were mostly taken form Financials for the year 2006, not for the year 2004). This way all subsequent corrections of errors and omissions were recognised. Such steps were essential in case of Shell which had to admit major overestimation of its reserves[15] and restate 2002–2005 financial statements.

Below presented table shows comparison between SMOG value and premium of Market Value over Book Value for the analysed companies.

Table 4.1. SMOG vs. Market Value-Basic comparison

Company	Year	MV[16]	BV[17]	SMOG[18]	MV-BV	SMOG as % of (MV-BV)	P.Y. SMOG[19] as % of (MV-BV)
		mo USD					
ChevronTexaco	2006	158 118	68 942	92 354	89 177	104	125
ChevronTexaco	2005	125 945	62 673	111 056	63 272	176	100
ChevronTexaco	2004	110 645	45 240	63 054	65 405	96	98
ChevronTexaco	2003	92 353	36 287	63 923	56 066	114	109
ChevronTexaco	2002	70 539	31 609	61 191	38 930	157	77
ChevronTexaco	2001	95 634	33 959	30 144	61 675	49	107
ChevronTexaco	2000	89 902	33 367	65 988	56 535	117	
BP	2006	218 192	86 517	90 600	131 675	69	97
BP	2005	221 099	85 147	128 200	135 952	94	65
BP	2004	209 520	85 092	88 500	124 428	71	65
BP	2003	181 958	79 167	80 500	102 791	78	74

[12]The prestigious 500 Fortune based on 2005 revenues listed ExxonMobil as the 1st, Royal Dutch Shell as the 3rd, BP as the 4th and Chevron as the 6th top global companies, the next one belonging to the same industry cluster: ConocoPhilips ranked the 10th.

[13]Or at least one of their "ancestors".

[14]Is some data were missing there other official company's sources were searched, e.g. 10-K Form.

[15]Using the most polite description for what happened in this corporation.

[16]MV = Market value as reported by a company for an year end.

[17]BV = Value of shareholders equity after minority interests stated in financials.

[18]SMOG as reported In financials.

[19]SMOG reported for the preceeding year against the current year end BV.

Table 4.1. Cont'd

Company	Year	MV[20]	BV[21]	SMOG[22]	MV-BV	SMOG as % of (MV-BV)	P.Y. SMOG[23] as % of (MV-BV)
		mo USD					
BP	2002	151 615	66 636	76 500	84 979	90	52
BP	2001	173 916	62 322	44 500	111 594	40	
BP	2000	172 671	65 554	N/A	N/A	N/A	N/A
ExxonMobil	2006	438 990	113 844	130 248	325 146	40	51
ExxonMobil	2005	344 491	111 186	164 307	233 305	70	47
ExxonMobil	2004	328 128	101 756	110 696	226 372	49	44
ExxonMobil	2003	269 294	89 915	99 240	179 379	55	54
ExxonMobil	2002	234 101	74 597	96 559	159 504	61	33
ExxonMobil	2001	267 577	73 161	53 248	194 416	27	50
ExxonMobil	2000	301 239	70 757	97 252	230 482	42	
Royal Dutch Shell	2006	222 945	114 945	53 797	108 000	50	67
Royal Dutch Shell	2005	200 614	97 224	72 295	103 390	70	54
Royal Dutch Shell	2004	193 720	84 576	55 985	109 144	51	54
Royal Dutch Shell	2003	178 169	72 848	59 451	105 321	56	57
Royal Dutch Shell	2002	151 123	60 444	60 362	90 679	67	45
Royal Dutch Shell	2001	140 889	56 160	40 414	84 729	48	80
Royal Dutch Shell	2000	186 872	57 086	67 906	129 786	52	

The following interesting observations could be made based on the aforementioned values:
- quite consistent results were obtained when the preceding year SMOG is compared to the premium of MV over BV, which is not the case if current values are examined for both – there is a tempting hypothesis that investors do know and take into account the SMOG for a preceding year rather than for a current year while taking investment decisions;
- all companies had been quoted well above their book value but, with exception of ChevronTexaco, not above the SMOG premium over book value;
- the SMOG value is greater (expressed as a percentage of MV/BV premium) for the couple renowned for its upstream activity than for the ones relying more on downstream (see Table 0-1 for reference).

---

[20]MV = Market value as reported by a company for an year end.

[21]BV = Value of shareholders equity after minority interests stated in financials.

[22]SMOG as reported In financials.

[23]SMOG reported for the preceding year against the current year end BV.

Table 0-1. Importance of upstream activities measured by upstream contribution to profitability

Company	Year	Comments	Total	Upstream	Upstream share in total
ChevronTexaco	2006	US GAAP, after interest and tax	17 138	13 142	77%
ChevronTexaco	2005	US GAAP, after interest and tax	14 099	11 724	83%
ChevronTexaco	2004	US GAAP, after interest and tax	13 328	9490	71%
ChevronTexaco	2003	US GAAP, after interest and tax	7230	6198	86%
ChevronTexaco	2002	US GAAP, after interest and tax	1132	4556	402%
ChevronTexaco	2001	US GAAP, after interest and tax	3288	4312	131%
ChevronTexaco	2000	US GAAP, after interest and tax	6322	6264	99%
BP	2006	IFRS	35 411	29 647	84%
BP	2005	IFRS	30 038	25 485	85%
BP	2004	IFRS	24 385	18 075	74%
BP	2003	IFRS	18 724	15 081	81%
BP	2002	IFRS	11 200	8277	74%
BP	2001	UK GAAP	19 608	14 498	74%
BP	2000	UK GAAP	21 253	15 710	74%
ExxonMobil	2006	US GAAP, after interest and tax	39 500	26 230	66%
ExxonMobil	2005	US GAAP, after interest and tax	36 130	24 349	67%
ExxonMobil	2004	US GAAP, after interest and tax	25 330	16 675	66%
ExxonMobil	2003	US GAAP, after interest and tax	21 510	14 502	67%
ExxonMobil	2002	US GAAP, after interest and tax	11 460	9598	84%
ExxonMobil	2001	US GAAP, after interest and tax	15 320	10 736	70%
ExxonMobil	2000				
Royal Dutch Shell	2006	US GAAP, after interest and tax	37 678	29 377	78%
Royal Dutch Shell	2005	US GAAP, after interest and tax	37 341	25 268	68%
Royal Dutch Shell	2004	US GAAP, after interest and tax	26 220	17 335	66%
Royal Dutch Shell	2003	US GAAP, after interest and tax	12 496	9105	73%
Royal Dutch Shell	2002	US GAAP, after interest and tax	9722	6796	70%
Royal Dutch Shell	2001	US GAAP, after interest and tax	10 350	7963	77%
Royal Dutch Shell	2000	US GAAP, after interest and tax	12 719	10 059	79%

The above presented review is definitely not sufficient to establish a clear link between SMOG and a value of oil and gas company perceived by the market. But it gives a strong indicator that such measure may carry valuable information to investors.

The SMOG value was compared to capitalised costs of exploration and development activities. It was expected that the former one will be significantly higher than the latter since a profit oriented company shall guide its activities in such way that expected returns should prevail over costs incurred. It is tempting to assume that capitalised costs to SMOG ratio reflects efficiency of exploration and development activities.

Table 0-2. Relation between E&D costs and SMOG

Company	Year	Comments	Oil and Gas Reserves		Capitalised E&D costs	SMOG	Capitalised costs as % of SMOG	SMOG Premium
			Oil (mo barrels)	Natural gas (mo sq ft)	mo USD			over cap. E&D costs
ChevronTexaco	2006		7506	22 884	58 003	92 354	63	34 351
ChevronTexaco	2005		8000	23 423	53 597	111 056	48	57 459
ChevronTexaco	2004		7973	19 675	35 116	63 054	56	27 938
ChevronTexaco	2003		8599	20 191	34 914	63 923	55	29 009
ChevronTexaco	2002		8668	19 335	32 038	61 191	52	29 153
ChevronTexaco	2001		8524	19 410	30 457	30 144	101	−313
ChevronTexaco	2000		8519	17 844	30 484	65 988	46	35 504
BP	2006		9781	45 931	60 906	90 600	67	29 694
BP	2005		9565	48 304	55 977	128 200	44	72 223
BP	2004		9934	48 507	53 459	88 500	60	35 041
BP	2003		10 081	48 024	50 975	80 500	63	29 525
BP	2002		9165	48 789	53 125	76 500	69	23 375
BP	2001		8376	46 175	50 740	44 500	114	−6240
BP	2000		7643	43 918	48 745	N/A	N/A	N/A
ExxonMobil	2006		11 568	67 560	70 182	130 248	54	60 066
ExxonMobil	2005		11 229	66 907	63 761	164 307	39	100 546
ExxonMobil	2004		11 651	60 362	62 949	110 696	57	47 747
ExxonMobil	2003		12 856	54 769	59 875	99 240	60	39 365
ExxonMobil	2002		12 623	55 718	49 764	96 559	52	46 795
ExxonMobil	2001		12 312	55 946	44 733	53 248	84	8515
ExxonMobil	2000		12 171	55 866	44 253	97 252	46	52 999
Royal Dutch Shell	2006		4196	44 142	58 071	53 797	108	−4274
Royal Dutch Shell	2005	2005−2002 restated in 2006	4191	39 616	48 151	72 295	67	24 144
Royal Dutch Shell	2004	2005−2002 restated in 2006	4888	40 567	48 906	55 985	87	7079
Royal Dutch Shell	2003	2005−2002 restated in 2006	5814	41 558	46 463	59 451	78	12 988
Royal Dutch Shell	2002	2005−2002 restated in 2006	6640	41 065	44 010	60 362	73	16 352
Royal Dutch Shell	2001		6107	42 059	28 445	40 414	70	11 969
Royal Dutch Shell	2000		9751	56 283	25 184	67 906	37	42 722

The above presented results are consistent with general opinion on efficiency of the exploration and development activities run by the four companies. Three of them, with a notable exception of Shell, reported the E&D capitalised costs to SMOG ratio falling between 50 and 70% over almost all reviewed years. Shell, which in turn had enormous problems with such activities leading even to false reporting aimed at covering the real state of its upstream business, eventually reported the ratio above 100% and has to take appropriate steps in order to match others' performance.

CONCLUSIONS

Financial reporting of mining companies system lacks clear representation of their mineral assets not only in Poland but even in the leading mining countries. This creates one of the biggest challenges for regulators since importance of the mining industry can not be anyhow neglected. As in many other cases historical costs approach can not provide satisfactory information for none of the key stakeholders: owners, creditors and governments. Therefore, as in case of various other classes of assets, a new, at least supplementary, measure for mineral assets value shall be elaborated and implemented. The SMOG, despite its deficiencies and limited use, seems to be one of the obvious propositions.

REFERENCES

[1] Brealey R., Myers S. 1996: Principles of Corporate Finance. McGraw-Hill, New York.
[2] British Petroleum Annual Reports and Accounts 2001–2006.
[3] ChevronTexaco Annual Reports 2001–2006.
[4] Copeland T., Koller T., Murrin J. 1990: Valuation. Managing the Value of Companies. Wiley & Sons, New York.
[5] ExxonMobil Annual Reports 2001–2006. K-10 Forms 2001–2005, SAR 2004–2006.
[6] Johnston D., Bush J. 1998: International Oil Company Financial Management in No Technical Language. PennWell, Tulusa, USA.
[7] IAS and IFRS 2007.
[8] KPMG: Mining – A Survey of Global Reporting Trends 2003. KPMG International.
[9] KPMG: Global Mining Reporting Trends 2006. KPMG International.
[10] Royal Dutch Shell Annual Reports 2001–2006.
[11] Uberman R., Uberman R. 1997: Wybrane problemy wyceny wartości złóż kopalin eksploatowanych odkrywkowo (Selected Issues of Open-Cast Mineral Deposits Appraisal). Górnictwo Odkrywkowe, nr 3, Wrocław 1997.
[12] Uberman R., Uberman R. 2005: Wycena wartości złóż kopalin. Metody, problemy, praktyczne rozwiązania (Mineral Deposits Appraisal. Methods, Challenges, Hands-On Solutions). Wyd. AGH – Uczelniane Wydawnictwa Naukowo-Dydaktyczne, Kraków 2005.

International Mining Forum 2008, Sobczyk & Kicki (eds) © 2008 Taylor & Francis Group, London, ISBN 978-0-415-46126-9

Application of WTP and WTA Categories in Valuation of Natural Resources

Krystian Pera
University of Economics, Katowice, Poland

ABSTRACT: The main issue of this study concerns problems of the natural environment evaluation. The starting point of analysis undertaken is the theory of neoclassical economy. The instruments that enable to estimate the value mentioned, the Willingness to Pay (WTP) and the Willingness to Accept (WTA) were assumed. These instruments were introduced from Hicks's prosperity measure: Compensating Variation (CV) and Equivalent Variation (EV).

Twofold depiction of this issue was presented (analytical and graphical) in two possible cases: the improvement of environmental quality and deterioration of its condition.

Another analysed issue was the consumer surplus, which together with total expense on the achieving specific level of the environmental quality, can be treated as a measure of complete value of the environment.

In the final part of the study, the difficulties connected with the measurement of the surplus were presented and also four variations were described: Quantitative and Price Compensating and also Price and Quantity Equivalent.

KEYWORDS: Natural resources, Willingness to Pay, Willingness to Accept, valuation of natural resources

1. BACKGROUND

The issues related to valuation of natural resources are far from simple. There are numerous reasons behind their methodological difficulties.

The most important ones are:
- Differing perceptions of values for evaluating investment projects and for estimating the social value of a resource.
- A varied level of values depending on the time of resource exploitation and its introduction into the economic and financial turnover (the present value and the option value of a resource).
- Lack of valuation methodologies with a relevant degree of universality for these applications.
- Shortage of money as a universal measure of value.

Such a broad range of areas that need to be addressed, as preliminarily outlined, clearly implies that there are limitations to the precise valuation of resources available in the natural environment, and consequently attempts should be taken to try and find increasingly effective measures. Estimating values of such goods is neither simple nor unequivocal. The complexity of this problem results, first of all, from the lack of fully objective measures, similar to the ones which are used e.g. on the financial markets. The subject of valuation as such – natural resources – is a specific and unique target of economic research.

Its uniqueness, from the economic point of view, is mainly determined by the following causes:
- the timeless value of natural resources;
- resource depletion increasing over time;
- existence of non-utility values;
- quasi – public character of natural resources.

Out of the features listed above, the last one in particular is a source of methodical difficulties in valuation and the main subject of this paper. We should note that goods and services offered by the natural environment are not directly related to market transactions, and thereby there are no market-determined measures, in form of prices, for them. In the same way, valuation of every natural resource is relative. Moreover, valuation of the natural environment also has to take into consideration the fact that the value category per se is highly aggregated in this case. In addition to the so-called utility value, its full value is also composed of the option value (when certain values are ascribed to resources due to their possible use in the future) and the existence value (the value per se, the essence of which is the very existence of the natural environment). The problem of natural resources valuation is usually limited to the utility value only, but this study adopts the concept of the full value estimation.

Therefore, two valuation platforms should be identified:
1. Investment valuation of resources – with its most frequent methodological reflection in modified methods of discounted cash flow.
2. Non-investment valuation of resources – the essence of which is estimation of values of natural resources at all their levels (utility, option and existence values).

So far the neoclassical approach, based on the utility category, in particular the marginal utility as the measure of value, has been regarded as the best-developed concept or actually the only approach showing the features of the complete concept of non-investment valuation of natural resources.

The neoclassical method of environmental valuation, adopted here as the methodological basis, stems from:
- measuring of limitation;
- identification of individual and collective preferences;
- social measuring of welfare used to derive the aggregated demand for a particular environmental quality.

2. THESIS AND METHODICAL ASSUMPTIONS

In the paper, the author adopted the following thesis: utility measures, as understood in the neoclassical economics, constitute the appropriate methodological basis for valuation of natural resources. The instruments used to examine this value are the willingness to pay (WTP) and the willingness to accept (WTA) concepts, derived from so-called Hicksian welfare measures: the compensating variation (CV) and the equivalent variation (EV).

Consequently, the paper is based on the following methodical assumptions:
- full value of natural resources is relative and corresponds to the marginal utility level at a specific aggregated demand for environmental goods;
- the valuation concept is based on ordinal and cardinal utility measures in accordance to the principles accepted by the welfare economics;
- the reference point for the environmental valuation is the aggregated demand curve by Hicks (Friedman-Marshall), understood as demand reported every time at a maximum price that a consumer is willing to pay for the subsequent goods.

3. AN ANALYTIC APPROACH TO THE COMPENSATING VARIATION (CV) AND THE EQUIVALENT VARIATION (EV)

Let's assume the following output data:
1. [X] – vector of market goods;
2. [P] – vector of market-goods prices;
3. Q – scalar of environmental quality;
4. U – utility level[1].

With such output data, the initial position of the Y^0 consumer may be expressed analytically as:

$$Y^0 = f\{[X,P], Q^0, U^0\}.$$

I further assume that the environmental quality is subject to change (improvement or deterioration). Then, the change in the consumer's utility level (improvement or deterioration respectively) may be regarded as an equivalent measure of the change in the environmental quality.

Consequently, it is assumed that:
- $Q^1 > Q^0$ – the environmental quality improved;
- $Q^1 < Q^0$ – the environmental quality deteriorated.

Both cases should be analysed using the previously assumed measures in form of the compensating variation (CV) and the equivalent variation (EV). However, the compensating variation is an analysis of the change in the environmental quality without a possibility to return to the initial status, and the equivalent variation retains a possibility to verify the decision taken, and consequently return to the initial status.

The consumer's position after the change in the environmental quality, i.e. Y^1, can be formally presented as:

$$Y^1 = f\{[X,P], Q^1, U^1\}.$$

In this formulation, the compensating variation may be expressed as[2]:

$$CV = f\{[X,P], Q^1, U^1\} - f\{[X,P], Q^1, U^0\}.$$

If $CV > 0$ with an unchanged vector of [X,P] it means that $U^1 > U^0$, so the change in the environmental quality from Q^0 to Q^1 led to an increase in the consumer's utility. Therefore, the ($U^1 - U^0$) difference corresponds to the maximum amount of money that the consumer would be willing to pay for this improvement in the environmental quality from Q^0 to Q^1 so that the consumer's utility level could remain not lower than the initial status of U^0. This difference corresponds to the WTP value.

If $CV < 0$, then with an unchanged vector of [X,P] $U^1 < U^0$, which means that the change in the environmental quality from Q^0 to Q^1 resulted in a decrease in the consumer's utility. Consequently, a conclusion may be drawn that $Q^0 > Q^1$. The $U^1 - U^0$ difference corresponds to the minimum amount of money that the consumer would be wiling to accept in return for the deterioration in the environmental quality so that the consumer's utility level could remain not lower than the initial status of U^0. This difference is the WTA measure.

---

[1] Every time the 0 subscript denotes the environmental status before the changes, and the 1 subscript after the changes respectively.

[2] In the equation the utility level of U changes and refers to a higher environmental quality of Q^1. In that way the assumption stating that the change in the environmental quality is valued by the change in the utility level is executed.

The equivalent variation, in turn, may be expressed as:

$$EV = f\{[X,P], Q^0, U^1\} - f\{[X,P], Q^0, U^0\}.$$

If $EV > 0$, then with an unchanged vector of $[X,P]$ $U^1 > U^0$, which means that the utility level after the change in the environmental quality is higher than before the change, therefore $Q^1 > Q^0$. The $(U^1 - U^0)$ difference corresponds to the minimum amount of money that the consumer would be willing to accept for renouncing the improvement in the environmental quality so that the consumer's utility level could remain not lower than U^1, i.e. in the situation when there was a change in the environmental quality. This situation corresponds to the WTA value.

If $EV < 0$, then with an unchanged vector of $[X,P]$ $U^1 < U^0$, the environmental quality deteriorated from Q^0 to Q^1, so $Q^0 > Q^1$. The $U^1 - U^0$ difference corresponds to the maximum amount of money that the consumer would be willing to pay for the absence of deterioration in the environmental quality from Q^0 to Q^1 so that the consumer's utility level could remain not lower than U^1. This corresponds to WTP.

4. THE GRAPHIC PRESENTATION OF THE COMPENSATING VARIATION (CV) AND THE EQUIVALENT VARIATION (EV)

The valuation concept based on the WTP and WTA measures may be presented graphically by means of indifference curves, i.e. the utility theory in the ordinal approach or by means of the Marshallian demand curve, i.e. the utility theory in the cardinal approach.

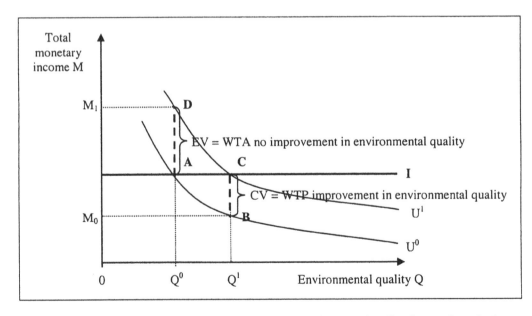

Figure 1. The WTP and WTA concepts in the improvement of environmental quality. Source: the author's own study based on [1], [4]. Q^0 – the level of environmental quality prior to the change; Q^1 – the level of environmental quality after the change; U^0 – the utility level prior to the change in the environmental quality; U^1 – the utility level after the change in the environmental quality

The resulting four possible variations may be analysed in the simplest way by examining the cases concerning the WTP and the WTA for the improvement in the environmental quality, which correspond to the compensating variation and the equivalent variation respectively, and after that the cases concerning the WTP and the WTA which correspond respectively to the equivalent variation (EV) and the compensating variation (CV) for the deterioration in the environmental quality from Q^0 to Q^1.

The equivalent variation (EV) and the compensating variation (CV) for the improvement in the environmental quality from Q^0 to Q^1 are illustrated by Figure 1. On the axis of ordinates there is the M good, which represents the total monetary income of an individual, by means of which the individual can buy any combination of market goods and services. On the axis of abscissas there is the environmental quality of Q. It is assumed that the environmental quality is a quasi-public good and may be provided on a free of charge basis only. In this case, the total income of the individual is allocated for market goods. The budget line of I is therefore parallel to the axis of abscissas.

By means of the compensating variation, it should be determined what a maximum amount the individual is willing to pay for the improvement in the environmental quality from Q^0 to Q^1 so that after the change the individual's utility level would not be worse than in the initial status of U^0. Let's assume that initially the individual is at point A on the indifference curve, representing the utility level of U^0. Following the improvement in the environmental quality, the individual would be positioned at point C on the indifference curve, representing the highest utility level of U^1, as the individual retains their initial monetary income with the higher environmental quality, and due to that the individual may enjoy all the market goods that may be purchased for this amount of money. In order to return to the initial status on the indifference curve, the individual has to sacrifice some of their income. This figure equals BC. After this amount is paid, the individual will have a lower disposable income of $0M_0$, but a better environmental quality of Q^1. This means that the individual will find themselves at a new point on the indifference curve of U^0 – at point B.

When using the equivalent variation, it should be determined what a minimum amount of money the individual is willing to accept in return for absence of improvement in the environmental quality from Q^0 to Q^1 so that the individual's utility level would not be lower than U^1. The contemplations here should start from point C on the indifference curve which represents the higher utility level of U^1. If the improvement in the environmental quality was desisted, the individual would be positioned at point A on the lower indifference curve of U^0. At this point the individual has a disposable income in its initial amount, but the environmental quality is lower and equals Q^0. In order to return to the higher indifference curve of U^1, the individual has to obtain a higher monetary income. This figure equals AD. After obtaining this amount, the individual will continue to have the environmental quality of Q^0, but they will also have a higher disposable income of $0M_1$. This means that the individual will find themselves at a new point on the indifference curve of U^1 – point D.

The cases concerning the WTP and WTA values, corresponding respectively to the equivalent variation (EV) and the compensating variation (CV) for the deterioration in the environmental quality from Q^0 to Q^1, are illustrated in Figure 2.

When using the compensating variation it should be determined what a minimum amount of money the individual is willing to accept for the deterioration in the environmental quality from Q^0 to Q^1 so that the individual's utility level after the change could remain not lower than the initial figure of U^0. Let's assume that the individual is initially positioned at point C on the indifference curve, representing a higher utility level of U^0 (the utility level prior to the change). After the deterioration of the environmental quality to the level of Q^1 the individual would find themselves at point A on the indifference curve representing a lower utility level of U^1, in which the individual has an initial disposable income but a worse environmental quality of Q^1. In order to return to the higher indifference curve, the individual has to obtain a higher monetary income. This additional in-come equals AD. After obtaining this amount, the individual will continue to have a lower environmental quail-ty of Q^1 but it will have a higher disposable income of $0M_1$ and will find themselves at a new point on the indifference curve of U^0 – at point D.

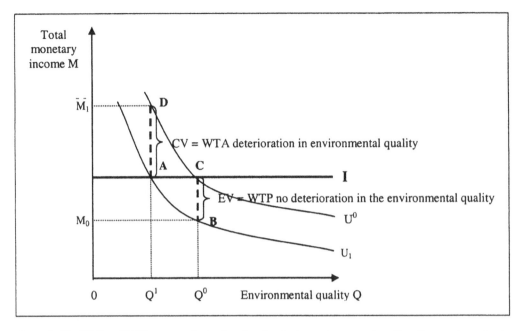

Figure 2. The WTP and WTA concept in the deterioration in the environmental quality. Source: the author's own study based on [1], [4]. Q^0 – the level of environmental quality prior to the change; Q^1 – the level of environmental quality after the change; U^0 – the utility level prior to the change in the environmental quality; U^1 – the utility level after the change in the environmental quality

When using the equivalent variation, it should be determined what a maximum amount the individual is willing to pay for absence of the deterioration of the environmental quality from the level of Q^0 to Q^1 so that the individual's utility level could remain not lower than U^1. The contemplations here should start from point A on the indifference curve which represents the lower utility level of U^1. If the deterioration in the environmental quality from Q^0 to Q^1 was desisted, the individual would be positioned at point C on the lower indifference curve of U^0.

At this point the individual retains its initial monetary income, and in addition to that, the individual can enjoy a better environmental quality of Q^0. In order to return to the lower indifference curve, the individual has to renounce part of its monetary income. This figure equals BC. After paying this amount, the individual will continue to have the higher environmental quality of Q^0, but it will also have a lower disposable income of $0M_0$ and the individual will find themselves at a new point on the indifference curve of U^1 – point B.

CONCLUSION

The paper presents an opinion concerning the measurement of the environmental quality, as well as a certain philosophy in perception of this issue. The problems, however, turn out to be quite complex in terms of methodology. The difficulties are not only related to the subject of the valuation but they, first of all, result from the very concept of the environmental value.

The starting point for the analysis is the neoclassical theory of utility. Such an approach means also that the value of the natural environment is estimated on the basis of the changes in the utility levels, and it is a relative valuation, i.e. the measure of the value is not only the price of a good but also the consumer's quantified willingness to pay for this good. As shown, both values may differ substantially.

The instruments used for estimation of this value are the Hicksian welfare measures, based on which the Willingness to Pay (WTP) and the Willingness to Accept (WTA) were explained. A two-fold depiction of this issue (analytic and graphic) is shown for two possible cases: improvement in the environmental quality and its deterioration.

The subsequent analysed issue is the consumer's surplus, which ·ι gether with the total monetary expenditure for achievement of a particular level of the environmental quality, may be regarded as a measure of the full value of the environment. The consumer's surplus is a manner for estimation of values that is relevant but quite difficult in terms of methodology. Therefore, the paper additionally presents the difficulties related to measuring of the surplus and describes the four variations: the quantitative and price compensating, as well as the price and volume equivalent.

REFERENCES

[1] Ahlheim M., Buchholz W.: WTP or WTA – Is that the Question? Reflections on the Difference Between "Willingness to Pay" and "Willingness to Accept". Wyd. Uniwersytetu w Regensburgu oraz Politechniki w Cottbus, 2002.
[2] Blaug M.: Teoria ekonomii. PWN, Warszawa 2000.
[3] Ekonomiczna wycena środowiska przyrodniczego. Praca zbiorowa pod red. G. Andersona i J. Śleszyńskiego, wyd. "Ekonomia i Środowisko", Białystok 1996.
[4] Pera K., Baron M.: Szacowanie pełnej wartości środowiska przyrodniczego. Ujęcie metodyczne. Wyd. "Ekonomia Środowisko", Nr 1/2002.
[5] Woś A.: Ekonomika odnawialnych zasobów przyrody. PWN, Warszawa 1995.

International Mining Forum 2008, Sobczyk & Kicki (eds) © 2008 Taylor & Francis Group, London, ISBN 978-0-415-46126-9

Strategy of Development and Value Creation in Mining Industry on Example of Lundin Mining Corporation

Arkadiusz Kustra
AGH – University of Science & Technology, Cracow, Poland

Krzysztof Kubacki
KGHM Lubin, Poland

ABSTRACT: Strategies of mining companies concentrates on building up of value creation through specific key drivers. One of them is development of deposits which mining companies may acquire by mergers and acquisition or by realizing their own mining projects.

Lundin Mining Corporation can serve as a perfect example of how fast end efficiently a mining company is able to grow and create value for its shareholders. In an industry characterized by long-term investments it has transformed from a small prospecting company to a diversified senior mining company within a very short time. It has done so with methodical approach to value creation and aggressive development strategy. Company's assets together with development projects and exploration portfolio enable Lundin to grow further and strengthen its position in the extractive industry.

INTRODUCTION

The fundamental objective of corporate management has become creation of its value for the sake of owners and other stakeholders concerned about functioning and development of a business activity. In the traditional approach the process of value creation may be identified on three levels: operational, investment and financial. According to Damodaran, the factors of value creation are those which pertain to increase of operational margin, maintenance of the rate of growth, optimization of capital costs and effective assessment rates as well as to investments in fixed assets and working capital. In addition to the above, enterprises choose the exogenous development through the consolidation process.

The present article deals with issues regarding the strategy of value creation in the mining industry which is characterized by the use of special techniques and technologies as well as mining and geological conditions which are deemed to constitute the natural environment in which mining companies usually exist. It discusses the representative factors of value creation and increasing trends of amalgamations and takeovers in the industry. The analyses seem to corroborate the fact that companies tend to concentrate on strategies which allow the increase of owned resources by means of development of the existing and brand new mining projects as well as reduction of costs of functioning and consolidation processes. The aforementioned trends resulted in the emergence of a specific structure of the mining sector worldwide, with three types of companies distinguished by strong domination of diversified corporations. In order not to leave analyses and conclusions unsupported by practical examples, let us consider the Lundin Mining company which growth fully reflects the strategies of market behaviour aiming at the increase of the mining companies' value worldwide.

FACTORS OF VALUE CREATION OF MINING COMPANIES

The mining companies are characterized by the specific scope of operations and activity due to applied technologies and the work organization systems unheard of in other types of enterprises. Furthermore, they demonstrate particular sensitivity to market environment, geographical conditions and price ruling in the places they function. The foregoing immediately results from impossibility to relocate extracted minerals and their inevitable depletion. According to A. Lisowski extraction of mineral resources may be compared to non-renewable storehouse generously provided by the nature and gradually emptied by men [2].

Impossibility of translocation of the extracted mineral resources accounts for great impact of legislative bodies and local communities on functioning of such mining companies in terms of social and environmental issues. Thus, the sources of value creation in the strategy of building up the mining company value must be analyzed on the basis of both financial and extra-financial benchmarks which explain their specific activity in the most comprehensive way.

The value of the mining company and prospects of building up such value in the future is, first of all, determined by anticipated period of extraction of mineral deposits. The aforesaid period is dependent on possessed and evidenced industrial resources destined for extraction in the future. The intensity of extraction of minerals as well as the rate of discovering new deposits is not to be neglected, too.

Recognition and evidencing of deposits of the mining plant is burdened with uncertainty in respect of an actual size of mineral deposits and their quality as defined by special quality parameters (e.g. calorific value, ash content, sulphur content, etc.)

They are identified within the framework of geological-exploratory research. The above-mentioned quality parameters of extraction determine the range and depth of enrichment processes as a result of which concentrates are obtained referred to as merchantable products fit to be offered for sale in external market.

In terms of the duration of growth as a factor which builds up the value of a mining company it might be regarded as an average period of mineral deposit exploitation, taking into consideration the size and quality of existing resources with their speed of exploitation established in advance.

The rate of increase of sales refers mainly to the sale prices which are one of the most vital external factors responsible for building up the value of the mining companies. Their level is determined by trends in Rock and Mineral Exchange worldwide, most often influenced by the law of supply and demand but also resulting from speculative operations undertaken by certain investors and active hedging funds. Similarly, seasonality and structure of sale of various types of minerals (in terms of quality) are not irrelevant to sales turnover. Furthermore, mining plants which are engaged in extraction of minerals may enhance the process by obtaining concentrates available at different prices. The level of sales turnover also depends on currency rates which have significant impact on effectiveness of output export and, in addition to the above, may play a vital role in setting the parity of prices for extracted and enriched raw materials as converted into actual or standard unit.

The detailed analysis of sales turnover combined with operational costs determines yet another factor building up value, i.e. operational profit margin. In case of mining companies deemed to be capital-intensive and cost-consuming such detailed analysis of operational costs may lead to reduction of idle production capacity and simplification of spatial model of making deposit available. The above-mentioned model determines the underground system of horizontal and vertical workings necessary to carry out the mining process. Accomplishment of such workings and their maintenance implies generation of fixed costs related to owned mining potential, which, in turn, suggests opportunities allowing flexible behaviour towards fluctuations occurring in the ever-changing economic situation. Liquidation of redundant workings, concentration of mining production and application of cheaper solutions in mining technology ensures the reduction of costs and thus affects the operational margin.

The mining industry is characterized by long-term investment cycles commencing from the geological research until the output in achieved as seen from the commercial point of view. This requires considerable investment expenditures made not only for the purpose of exploratory-geological works but also in respect of the creation of a spatial model of a mining company. Expenditures related to the aforesaid model of a mining company correlate with the value of explored resources only to a minor extent; on the other hand, however, they settle the business of any prospective exploration and existence of a mining plant. The permanent undertaking of exploratory works with the concurrent carrying out the extractive activities ensures the continuous excavation process.

Aside from investments in fixed assets essential for building up the value of a mining plant, there are also investments in working capital. This is to be understood as accumulating an appropriate amount of reserves and quick liabilities deemed to be material current assets under managed. Their high level accounts for pecuniary means freeze and generates additional demand for working capital, which, consequently, may induce the use of external sources of financing involving interest rates, which increases the financial risk of a company and ipso facto the cost of gaining capital. The reserve level in a mining plant is of great importance; this level may vary depending on the season of sales of particular minerals.

Another factor which affects the value of a mining company is the cost of the capital financing the current operations as well as investments. Cost is usually perceived as weighted average cost of capital (WACC) and it tends to be higher in the mining industry than in any other sectors. The condition described in the preceding sentence stems from the high risk of activity accompanied by hazards connected with extraction of owned mineral deposits. Moreover, mining plants, due to their specific character, tend to focus on long-term profits which is not always approved by investors and analysts who are mostly concerned with short-term profits. Therefore, their recommendation in financial markets may indirectly increase the cost of access to capitals. Analysts and investors also pay attention to lack of proper information and impossibility of performing reliable appraisal of mining companies' activity on the basis of existing financial statements. Measures calculated basing on the foregoing do not reflect all aspects of functioning and the specific character of mining plants which accounts for high level of investment risk and the increased rate of return.

To recapitulate, it is necessary to emphasize that the key sources of building up the value of mining companies are not only identified within the traditional areas of activity: operational, investment-based and financial but also involve extra financial factors related to specific activity of mining companies, i.e. political, social and environmental. The proper value reporting, taking into account all the above-mentioned sources and areas of creation, may contribute to appropriate appraisal of mining plants in financial markets.

STRATEGY OF GROWTH OF THE VALUE OF MINING COMPANIES BASED ON MERGERS AND TAKEOVERS

The present status of the mining industry, which seems to be dominated by large and diversified coal companies, is the consequence of the implemented strategic policy including:
- diversification of income and results according to the type of activity and the scope of operations, with the risk of activity and the costs of functioning reduced;
- increase of financial regime;
- more opportunities to gain capital to finance in particular high-risk investments;
- increase of value for owners.

Implementation of the above strategic factors suggests the increasing wave of mergers and takeovers in the mining industry. The consolidation processes ensured operational and financial synergy. The basic reason for mergers and takeovers of mining companies is diversification of activity, i.e. differentiation of revenues and results with regard to the structure of extraction process and geographical localization. The aim of the foregoing is to reduce the risk arising from periodic changes in metals and raw materials' prices at the exchanges worldwide as well as the political risk. The possibility of subsidizing respective segments of activity increased the competitiveness of mining plants.

It is also worth mentioning that mergers and takeovers allowed to multiply the mineral and metallic deposits owned by different subjects. Thus, the enterprises extend the life cycle of an organization without incurring redundant risk related to geological research and exploration of new deposits. Bearing in mind obligate expenditures to launch a mining plant and the long period of return of invested monies, mergers and takeovers may be deemed to constitute an attractive alternative to choose from to enhance growth. This is confirmed by the strategy adopted by the most powerful mining plants worldwide which constantly decrease the ratio of geological and mining research expenditures to the value of carried out mergers and takeovers.

Furthermore, consolidation improved optimization of procurement, enhanced competitiveness and strengthened the market standing in relations with contracting parties. On the side of financial effects one should appreciate wider access to capital and sources of financing as well as reduction of financial costs connected with debt service.

At present, when the prices of raw materials and metals are on the increase, the value of mergers and takeovers may be subject to re-evaluation with reference to the real value of combined assets. Likewise, it must be emphasized that the strategic character of mergers and takeovers is rather short-term since it does not contribute towards development of new mining technologies and exploration of new deposits. On the other hand, amalgamated companies are surely in a better position which allows to ensure easier and cheaper access to sources of financing the expensive geological and mining works.

Processes such as mergers and takeovers do increase the value of mining companies, which is also connected with sustainable development within which framework the mining business should be made profitable with the concurrent social acceptance and respecting the standards and provisions relating to environment protection.

Sustainable development of mining companies has become one of the major corporate strategies which implies that running an activity is much more than just a business, since it should also assure the economic development of a region in which mining companies operate. Such policy must generate more costs which only large, diversified corporations of long-market standing may choose to bear.

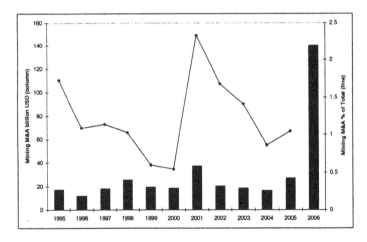

Figure 1: Mergers and Acquisitions in Mining Industry, 1995–2006.
Source: Brett D., Ericsson M.: M & A Survey 2007, Raw Materials Group, 2007

According to Fitch agency, the key motives for conducting mergers and takeovers in the mining industry are [4]:
- increase of geographical, product and market diversification aiming at reduction of operational risk;
- extension of the scope of operations and improvement of market standing and competitiveness;
- focusing on research and development;
- taking advantage of the economics of scale and synergic effects;
intensification of financial standing securing easier access to capital and sources of financing.

Table 1. The largest mergers and acquisitions since 1995

	Buyer	Target	Sector	Year	Amount [mln USD]
1.	Rusal – Russky Aluminii	SUAL Group	Aluminium	2006	30 000
2.	Freeport McMoran Copper & Gold Inc.	Phelps Dodge Corp.	Copper	2006	25 900
3.	Cia Vale do Rio Doce	Inco Ltd.	Nickel	2006	17 895.7
4.	Xstrata plc.	Falconbridge Ltd.	Diverse	2006	14 483.8
5.	BHP Billiton plc.	BHP Billiton Ltd.	Diverse	2001	14 000
6.	Anglo American plc.	De Beers Consolidated Mines Ltd.	Diamond	2001	11 440
7.	Barrick Gold Corp.	Placer Dome Inc.	Gold	2006	10 400
8.	Goldcorp Inc.	Glamis Gold Ltd.	Gold	2006	8600
9.	BHP Billiton Group	WMC Resources Ltd.	Diverse	2005	7300
10.	Aluminum Company of America	Reynolds Metals Co.	Aluminium	1999	4600
11.	Alcan Aluminium Ltd.	Alusuisse Group Ltd.	Aluminium	2000	4400
12.	RTZ Corporation plc.	Rio Tinto Ltd.	Diverse	1995	4000
13.	Alcan Inc.	Pechiney	Aluminium	2003	3841.4
14.	Aluminum Company of America	Alumax Inc.	Aluminium	1998	3800
15.	Anglo American Corp of South Africa Ltd.	Minorco SA	Diverse	1998	3699.9
16.	Texas Pacific Group Ventures	Aleris International Inc.	Aluminium	2006	3300
17.	Inco Ltd.	Voisey's Bay Nickel Co. Ltd.	Nickel	1995	3278.2
18.	Valepar AS	Cia Vale do Rio Doce	Diverse	1997	3150
19.	Anglogold Ltd.	Gold Assets	Gold	1998	3104.3
20.	Iamgold Corp.	Cambior Inc.	Gold	2006	3000

Source: Brett D., Ericsson M.: M & A Survey 2007, Raw Materials Group, 2007

STRUCTURE OF THE MINING INDUSTRY
AS A RESULT OF MERGERS AND TAKEOVERS

Consolidation occurring in the mining industry all around the world in responsible for development of a specific structure of the mining companies' market.

Detailed analyses indicate that the following may be identified within the market:
- senior mining companies,
- medium mining companies,
- junior mining companies.

According to data collected by the Raw Materials Group which is engaged in the analysis of raw materials' market, there are approximately 4000 various mining companies operating worldwide (aside from gas and crude oil mining enterprises).

According to the authors of the aforesaid analysis, the first group of mining companies which includes only 149 plants represents 60% of the value of the mining market worldwide. The medium mining companies group (957 plants) – represents 40% of the market value and the junior mining companies group, although the biggest in number (3067 plants) does not contribute to the value of the mining market worldwide since these companies are not involved in extraction and processing.

The first group, i.e. the senior companies, encompasses the mining companies for which operational activity constitutes the sale of highly processed raw materials acquired as a result of the value creation extension process, from the stage of extraction, through enrichment to refining of products. Similarly, such plants are diversified in terms of products and geographical localization, which implies that they are dedicated to exploitation of various minerals on different continents. Thus, they manage to reduce the risk of changes in mineral resources' prices which are characterized by specific cycles dictated by the existing state of affairs, and also reduce the political and legal risk related to conducting a business activity in developing and unstable economies.

The group to which medium mining companies belong is mostly engaged in realization of the extraction process. As a rule they own several mines in which mining of a particular type of mineral is carried out. Such companies do not represent product and geographical diversification.

The last group consists of small-sized companies for which the basic scope of operations includes exploration, discovering and making deposits available which, at a later stage, shall be exploited on an industrial scale. It is important to highlight that the extraction process itself shall most probably be undertaken by senior or medium mining companies who shall have purchased officially ascertained deposits. In practice, a part of companies referred to in this paragraph may own single mines and carry out the mining process in order to gain additional income. It is beyond doubt, however, that assets owned by such companies (understood as ascertained deposits and proper permits and licenses for their authorized extraction) constitute a potential target to be taken over and acquired by the biggest mining companies.

LUNDIN MINING CORPORATION

Lundin Mining Corporation is a Canadian based international mining and exploration company. It is involved in the extraction, development, acquisition and discovery of base metal deposits. The Company has six operating mines in Portugal, Sweden, Ireland and Spain, and approximately 2,000 employees. Lundin is a diversified, mid-tier copper, zinc, lead, nickel, and silver producer. Head office of the company is registered in Vancouver, Canada, while executive and operational offices are currently in Stockholm, Sweden. The Company intends to relocate its executive office to Geneva, Switzerland. Lundin Mining shares are listed on the Toronto Stock Exchange ("LUN") and the NYSE ("LMC"), and its SDRs are listed on the OMX Nordic Exchange ("LUMI").

Lundin Mining was incorporated under the name "South Atlantic Diamonds Corp." on September 9.1994. The Company changed its name to "South Atlantic Resources Ltd." on July 30.1996. In connection with a one-for-six share consolidation that took effect on April 2.2002, the Company changed its name to "South Atlantic Ventures Ltd." on March 25.2002. Later its common shares were listed on the Toronto Stock Exchange effective August 12.2004, and the Company changed its name to "Lundin Mining Corporation".

The Company owns six operating mines:
1. Neves-Corvo in Portugal, which is an underground copper and zinc mine located 220 kilometres southeast of Lisbon in the Alentejo district of Portugal, in the Iberian Pyrite Belt. Neves-Corvo in 2006 produced 78,576 tonnes of copper, 7,505 tonnes of zinc, and 645,521 ounces of silver, all metals contained in concentrates.

2. Aljustrel in Portugal, which is an underground zinc, lead and silver mine located approximately 40 kilometres northwest of the Neves-Corvo mine in Portugal. Aljustrel commenced production in December 2007. The mine is planned to reach full production in 2009 and produce 80,000 tonnes of contained zinc, 17,000 tonnes of contained lead, and 1.25 million ounces of silver.
3. Zinkgruvan in central Sweden. Zinkgruvan has been in production continuously since 1857. The primary metal produced is zinc, with lead and silver as by-products. It produced 75,909 tonnes of contained zinc, 31,850 tonnes of contained lead and 1.76 million ounces of silver in 2006.
4. Storliden in northern Sweden, which is an underground zinc and copper mine. It produced 27,824 tonnes of contained zinc and 10.642 tonnes of contained copper in 2006. The Storliden ore reserves are expected to be depleted during 2007 and the mine is scheduled for closure in the fourth quarter of that year.
5. Galmoy in Ireland, an underground zinc, lead and silver mine. Galmoy produced 60,055 tonnes of contained zinc, 13,256 tonnes of contained lead and 131,797 ounces of silver in 2006.
6. Aguablanca in Spain, an open pit nickel, copper and platinum mine. Aguablanca produced 14 million pounds of nickel in 2006.

In addition, Lundin Mining holds an extensive exploration portfolio, including interests in international ventures and development projects such as the world class Tenke Fungurume copper/cobalt project in the Democratic Republic of Congo, which is currently under construction and the Ozernoe zinc project under detailed feasibility study in Russia.

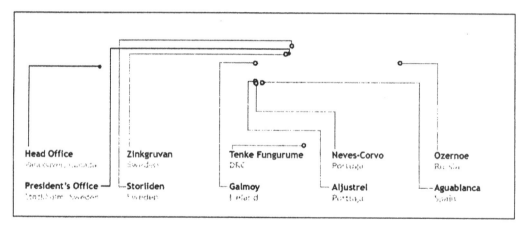

Figure 1. Location of Lundin Mining mines and mining projects. Source: www.lundinmining.com

LUNDIN MINING'S GROWTH STRATEGY

According to the mission of the Company the objective of Lundin Mining's operations is to create shareholder value by building and maintaining effective operations, while ensuring that the safety, environmental and social responsibilities of the Company are given primary focus. Furthermore when it comes to financial aspects the primary objective of Lundin Mining is to create value for its shareholders through profitable operations and growth. Value is generated by a combination of cash flow from producing assets and through exploration and prudent acquisitions. Cash flow from producing assets can be improved through professional management, leading to higher production levels and lower production costs. The goal of exploration is to add to the company's reserves and resources, while prudent acquisitions can improve cash flow as well as add reserves and resources. The company intends to fund acquisitions primarily through internally generated cash flow [1].

To create this value Lundin Mining Corporation has defined and implemented a simple and straightforward growth strategy.

It is based on the following principles, which were already mentioned in Company's financial objectives:
- Optimize performance of existing operations.
- Brownfield exploration near existing mines.
- Greenfield exploration in Portugal, Sweden, Spain and Ireland.
- Pursue opportunities for accretive acquisitions:
  - operational mines,
  - development projects,
  - exploration projects.
- Strategic investments in promising projects [2].

Until 2004 Lundin was an exploration company listed on the TSX-Venture. Ever since 2004 when the Company changed its name to Lundin Mining Corporation and its common shares were listed on the Toronto Stock Exchange it has very consistently and effectively worked towards reaching those strategic goals. It has enabled Lundin Mining to transform the Company from single commodity regional player to intermediate global diversified metals producer over a course of very short time.

Lundin has been constantly pursuing opportunities for accretive acquisitions. First of all it was looking for low cost operational mines which had a potential of being further developed.

Since that time Lundin has been able to acquire:
1. In June 2004 - Zinkgruvan zinc, lead and silver mine acquired from Rio Tinto. Lundin issued 20 million shares in May 2004 to raise a total of CAD 160 million for the transaction. Zinkgruvan is located about 200 km southwest of Stockholm, Sweden. The estimated mine life is at least 15 to 20 years. The mine is ranked in the lowest cost quartile among zinc mines in the world. Through the acquisition of Zinkgruvan Lundin transformed itself into a significant base metals producer with a clear European focus.
2. In January 2005 – Storliden copper and zinc mine located outside the town of Malå in the Skellefte Field in northern Sweden. North Atlantic Natural Resources AB (NAN) owned 100% of the Storliden mine. During most of 2004 Lundin held about 37% of the shares of NAN. In December 2004 it was decided Lundin would acquire Boliden AB's shares in NAN, resulting in Lundin holding a 74 % stake in NAN. Storliden is operated by Boliden. The transaction was financed from newly issued shares and cash coming from the sale of Zinkgruvan's silver production to Silver Wheaton. The remaining shares of NAN were also acquired by Lundin during 2005 resulting in full incorporation of NAN into Lundin. Storliden was put into production in 2002 with a relatively short mine life. Storliden acquisition strengthened Lundin's position of a base metals producer in Europe and added copper to commodity mix.
3. In May 2005 – Galmoy zinc and lead mine in Ireland. This acquisition was a result of a merger of Lundin Mining with ARCON International Resources which owned Galmoy. The transaction was a combination of cash offer and exchange of shares. The mine started production in 1997. It is a low cost mine. It has an expected Life of Mine of about 4 years, but it may be changed due to good near-mine exploration results. Galmoy acquisition resulted in Lundin's operations in new geographical location, and stronger position amongst base metals producers.
4. In November 2006 – Neves-Corvo copper and zinc mine in southern Portugal, following a merger with EuroZinc Mining Corporation. The transaction was executed by an exchange of shares. Neves Corvo has an estimated mine life until 2022. Merger with EuroZinc added another strong operation to Lundin's portfolio in a different geographical location. It also added significantly to Lundin's copper production.

5. In August 2007 – Aguablanca nickel and copper mine in south-western Spain, following the acquisition of Rio Narcea Gold Mines Ltd. This transaction was a cash offer, mixture of own cashflow and loan. It added nickel to Lundin's commodity mix.

All operational mines Lundin Mining was able to acquire were being evaluated and optimized to increase their performance. In the nearest future production levels at Neves-Corvo and Zinkgruvan will be significantly increased and their mine life extended.

Apart from producing mines the Company was looking for developmental mining projects with high potential that could be quickly brought into production.

Lundin acquired interest in the following:

1. Aljustrel Project in Portugal – acquired by Lundin in November 2006 as a result of a merger with EuroZinc. Located about 40 km northwest of Neves-Corvo. This zinc, lead, and silver mine commenced production in December 2007 after being on care and maintenance for 14 years. It has a mine life of at least 10 years and further exploration potential.
2. Tenke Fungurume Project in the Democratic Republic of Congo – Lundin Mining holds a 24.75% interest in this copper/cobalt project located in Katanga Province as a result of a merger with Tenke Mining Corporation in July 2007. The transaction was based on the exchange of shares. Freeport-McMoRan Copper & Gold Inc. holds a 57.75% interest and La Generale des Carrieres et des Mines, the DRC state mining company, holds the remaining 17.5% interest. Freeport is the operator of this project which is under construction and scheduled for production in 2009. Tenke Fungurume is considered the world's largest copper development with first phase production plans of 115,000 tonnes per annum copper and 8,000 tonnes per annum cobalt. Expansion plans would result in annual production of 400,000 tonnes of copper. It is a very low cost operation and less than half of the concession area had been explored to date.
3. Ozernoe Project in Russia – Lundin acquired from IFC Metropol in November 2006 a 49% interest in the Ozernoe project, a zinc/lead deposit located in the Republic of Buryatia in the Russian Federation for 125 million USD. A joint venture company was formed by both enterprises. A bankable feasibility study is currently underway and production is scheduled to start in 2010. The open-pit mine at full capacity is planned to produce annualy up to 350,000 tonnes of zinc and additional lead and silver concentrates.

Once these projects come into production it will significantly increase the levels of zinc, lead and copper produced by Lundin Mining. It will further improve the Company's cash-flow. Ozernoe and Tenke Fungurume mean new geographical mix of Lundin Mining's operations.

Since 2004 Lundin Mining was a fast growing producer of base metals. It first started however as an exploration company and continues to be very active in this field to stimulate its growth. In 2006 Lundin spent about 9,5 million USD on exploration works and employed 20 people in this division.

The Company focused on exploration activities it the following areas:
– brownfield in-mine and near-mine exploration concerning all existing operations with a view to expand reserves and resources, extend the mine life, etc.;
– greenfield exploration: in the Skellefte and Bergslagen districts of Sweden, Iberian Pyrite Belt in Portugal and Spain, Toral Project in Northwestern Spain, and Ireland, all in order to grow by finding new prospects, evaluating them, and developing exploration projects into commercial production.

In addition to investing in development projects and carrying out own exploration works Lundin Mining emphasizes the importance of acquiring interests in, and providing financial and technical support to, successful junior exploration and prospecting companies. Such strategic partnerships with these companies enable Lundin to participate in rewarding new exploration opportunities and discoveries. For Lundin it is yet another way to grow the business.

Since 2005 the Company has formed strategic partnerships with different exploration companies:

1. In August 2005 – acquisition of 19.9% (for approximately 3.4 million U$D) of Union Resources Ltd, an Australian exploration company that in 2000 discovered one of the world's largest known undeveloped resources of zinc, the Mehdiabad project in central Iran. Union Resources will own 50% of the project upon its completion. Mehdiabad also contains lead and silver, and is expected to produce around 500,000 tonnes of zinc annually.

2. In January 2006 – acquisition of 10% (for approximately 4.5 million USD) of Sunridge Gold Corporation, a Canadian exploration company. Sunridge has several advanced copper-zinc-gold exploration projects in Eritrea (northeastern Africa). In May 2007 Lundin invested further 6.75 million USD in Sunridge to hold a total of 19.6% shares.

3. In November 2006 – acquisition of 10% (for approximately 2.87 million USD) of Mantle Resources Inc., a Canadian exploration company focused on exploring the Akie zinc-lead project located in north-eastern British Columbia, Canada.

4. In January 2007 – acquisition of 14% (for approximately 2.6 million CAD) of Sanu Resources Ltd. a Canadian exploration company holding several base metals and gold properties in Eritrea, Burkina Faso, and Morocco.

5. In August 2007 – acquisition of 20% of Chariot Resources Ltd. (through the acquisition of Rio Narcea Gold Mines Ltd. which previously owned the shares in Chariot Resources) developing its 70% owned Marcona Copper Project in Peru, scheduled for production in 2009.

VALUE CREATION RELATED TO LUNDIN MINING GROWTH STRATEGY

The primary objective of Lundin Mining is to create value for its shareholders. The value of the Company listed on the stock exchange is reflected in its share price and the quoted market value. Since 2004 when Lundin Mining's shares were listed on the Toronto Stock Exchange and the Company was transformed from a junior exploration company, through small regional metals producer to a global diversified base metals producer its share price and market value have started to rise very rapidly. It obviously occurred during the time when all base metals prices experienced strong growth, which was very rewarding for all miners globally. Nevertheless it was too a result of the aggressive development strategy the Company was implementing.

Since 2002 market capitalization of Lundin Mining has risen from 10.5 million CAD to over 6 billion CAD during peak period in July 2007. With such a long list of development projects in its portfolio the Company's share price has a lot of upside potential.

However, value of the Company is not solely limited to current share price, as these rise and fall over periods of time. Like other mining companies Lundin creates value for its shareholders with a long-term vision.

This value is created in the following areas:
– people,
– world class assets,
– the Lundin Method,
– project pipeline,
– range of growth opportunities.

Lundin profits from having very talented employees who operate the mines and carry out exploration works. It is in the producing mines and successful exploration activities, where returns for shareholders are generated. Furthermore the Company is run by outstanding senior operating management team capable of maintaining the costs at required level. These two groups lay the foundation for successful value creation at Lundin Mining. It is worth mentioning that out of 8 persons in the senior management 3 are solely devoted to development of the Company (exploration, projects, and strategic partnerships).

People are followed by the world class assets. These are the operational mines where different base metals contained in concentrates and other by-products are produced. Company's cash flow stems mostly from sales of concentrates. In other words it is generated in the mines. All operations are being optimized and improved in order to reach maximum performance levels.

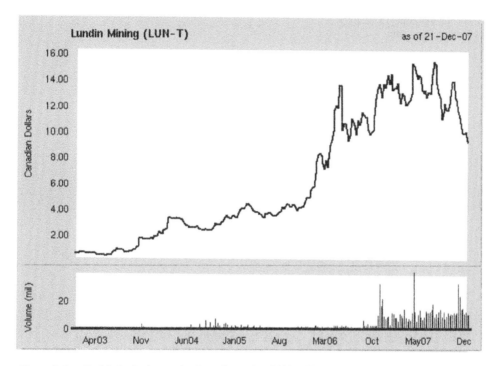

Figure 2. Lundin Mining's share price from December 2002 to December 2007. Source: www.tsx.com

The Lundin Method could be interpreted as the Company's way of running the business to successfully create shareholder value.

It is based on:

- financial strength which enables Lundin Mining to pursue accretive acquisitions;
- geographical mix of operations and projects which means that the Company does not limit itself in this field but intends to get involved in any given project as long as it is economically viable;
- talented employees from all over the world who can contribute to the Company's success;
- recognition of opportunities for the Company (exploration, development projects, strategic partnerships, etc.) that will enhance current and future economic value;
- entrepreneurial environment within the Company which allows for quick decision making and action being taken.

Project pipeline was thoroughly described in the previous chapter. It includes all production assets, and development assets which are all regarded as expandable. Further it includes exploration assets in different countries and as addition strategic investments in promising junior prospecting companies.

Range of growth opportunities stands on top of all previously mentioned areas. It is a mixture of internal opportunities (like performance optimization and in-mine exploration) and external growth (through greenfield exploration, strategic partnerships, acquisitions, etc.).

REFERENCES

[1] Brett D., Ericsson M.: M & A Survey 2007, Raw Materials Group, 2007.
[2] Lisowski A.: Podstawy ekonomicznej efektywności podziemnej eksploatacji złóż. GIG Publishing House, PWN Publishing House, Katowice–Warszawa, 30 p.
[3] Lundin Mining 2006 Annual Report.
[4] www.fitchratings.com (15.12.2005).
[5] www.lundinmining.com

International Mining Forum 2008, Sobczyk & Kicki (eds) © 2008 Taylor & Francis Group, London, ISBN 978-0-415-46126-9

Assessing Market Attractiveness of Power Generation Industry Companies for Hard Coal Producers

Paweł Bogacz

AGH – University of Science & Technology,
The Faculty of Industrial Economics and Management, Cracow, Poland

ABSTRACT: The paper presents the construction of a method used for detailed survey of power and power-and-heat generation plants' market attractiveness, which in the author's opinion could become a basis for creating marketing strategies of coal mines in Poland. In the construction of this method the relationship marketing idea was used and the analytical process presented was based on multidimensional comparison analysis and expert analysis tools. To illustrate the proposed method the paper contains an example based on a survey of the power generation industry and Kompania Węglowa S.A.

KEYWORDS: Marketing, market attractiveness, managing customer services, power generation industry, power plant, power-and-heat generation plant, mining company

1. BACKGROUND

Growing supply of industrial products observed for the past few decades, increasing globalization and the resultant competition forced many companies to undertake actions aimed at defending their market positions. To this end two lines of action were developed: restructuring and intensification of marketing activities.

Changes within a company involve mostly technical, financial and organisational restructuring (described in numerous publications). Its goal is to increase effectiveness and create clear organisational structure allowing for reliable and fast exchange of information. The result of such changes can be summarised as increasing effectiveness of managing the company's mineral reserves.

The second approach is directed towards a deep penetration of the market and a thorough search for an attractive customer group. A properly targeted marketing results in increased sales, whereas full control over marketing expenses ensures increase in profitability of the enterprise.

Within the group of market-related activities competition forced companies to change the philosophy in the way they operated from the initial manufacture-sell (transactional) to relationship.

The situation depicted above has been, for the past twenty-odd years, valid for Poland as well. Many branches of industry managed to make up for the decades of neglect relatively quickly. There are some, however, where the defunct manufacture-sell way of managerial thinking still prevails. One of them is mining, hard coal mining in particular. In so far as the sector has since 1992 been subject to wide-spread restructuring (internal change), its marketing activities still need to be intensified and broadened. A survey conducted in 2005 showed that application of modern marketing concepts in mining companies [1] remained at a very low level. This proved the necessity to raise and tackle this problem.

Taking into account the promising results of the reforms conducted in the mining sector, further in the paper the author proposes a research algorithm which is in line with the concept of relationship marketing and may serve mining companies as an additional tool allowing for detailed analysis of their largest customer group, i.e. power and heat-and-power generation plants. Within the algorithm, focusing particularly on the segmentation (grouping) stage, a system for conducting an assessment of market attractiveness of this customer group was proposed.

2. A FEW REMARKS ON MARKET ATTRACTIVENESS

The issue of assessing market attractiveness of customers for their suppliers is one of the newest problems of marketing methodology.

According to the latest and most comprehensive definition given by Cheverton [2] market attractiveness means "A set of multicriterion characteristics depicting a customer, which provide information on sales volume that could possibly be reached in business contacts with this company".

A large share of the effort at defining the idea of market attractiveness comes from Polish economists. Let's quote definitions given by Mazurek-Łopacińska [3]: "A measure to what extent a customer meets the needs of supplier (high sales volumes, high mark-up or sales volume low as yet but high potential)" and by Chodorowska and Krokosz [4]: "A set of variables characterizing the customer from the point of view of his buying potential".

The need to assess market attractiveness of consumers for their suppliers is one of the principal elements of relationship marketing. Together with value marketing the concept is one of the most recently added to marketing activities of companies. Because of its links to the process of market attractiveness assessment, basic principles of relationship marketing were given below.

The concept of relationship marketing revolutionized the way market was viewed and the supplier's position in contacts with customers built. With development of information society and the so called knowledge-based economy it turned out that proposed by Culliton in 1948 and further enhanced by Borden [5] in 1964 concept of transaction marketing did not fully describe the essence of building solid and profitable ties with customers. This led to invention of the concept of relationship marketing, also called partnership marketing.

Gronroos [6], who is considered its creator, defines it as: "A profitable structure, maintaining and developing ties with customers and other partners while fulfilling the goals of both sides through exchange of values and meeting obligations".

Figure 1. The process of relationship marketing

The concept found its first application in the service industry but was adapted to the specifics of the manufacturing industry already in 1983 (by IBM) (after [7]). Development of relationship marketing theory led to broadening of the definition of the concept to encompass the phrase: "Understanding and anticipation of customers' needs, integration of resources, means and actions of the company to profitably and effectively provide products and services in a manner which is more efficient than that of the competition" [8], as well as the final 5Is (Identification, Individualization, Interaction, Integration, Integrity) [9].

Within relationship marketing three main sub-processes (stages) can be identified, as is schematically shown in Figure 1.

By using market research, segmenting it into profitable sectors (objects), developing differentiated marketing strategies and controlling the effectiveness of the ac᛫ ns taken (Fig. 1), relationship marketing enables to create long-lasting and profitable relationships with customers.

According to this approach the customer is asked to specify his needs while at the same time his market attractiveness for the company is assessed. The findings are used to construct a differentiated system of marketing (targeted at sectors, customer groups) and after its introduction controls are used to measure its effectiveness.

3. ALGORITHM FOR ASSESSING MARKET ATTRACTIVENESS OF COMPANIES IN THE POWER GENERATION INDUSTRY

When developing the method the author mainly used multicriterion comparison analysis and expert analysis tools. The algorithm of the proposed analytical system is schematically shown in Figure 2.

The first part of the method comprises assumptions and initial design. At this stage the variables that in a multifaceted manner characterize the power-generation company are defined. A comprehensive description of companies is essential for operational market segmentation in which the author conducted his research.

The author decided to propose that analysing power and power-and-heat generation plants took into account 77 variables, responsible for their four potentials:
- Production potential,
- Sales potential,
- Financial potential,
- Environmental potential.

It was not possible to present all the variables in this paper due to the limited space, their detailed descriptions can be found in [10]. It must be stressed, however, that the variables were selected by the author based on his years-long experience and observations of the power generation industry and that financial variables are concordant with the International Accounting Standards. The set of variables was also approved by two largest Polish agencies for statistics and energy market monitoring, i.e. Agencja Rynku Energii S.A. (the Energy Market Agency) and Urząd Regulacji Energetyki (the Energy Regulatory Office). It should be also said that one of the criteria for selecting the variables was availability of relevant data.

As the second part of the algorithm data collection and verification was assumed. All information necessary to calculate variables for all power and heat-and-power generation plants in Poland could be obtained from Biuletyn Informacji Publicznej [11] and Agencja Rynku Energii S.A.

After Draper and Smith [12] the data collected was verified for coincidence with normal distribution with Kolmogorow's λ test, Shapiro-Wilk test and by compiling basic descriptive characteristics of each variable. Every such examination should be followed by a decision to accept a variable as suitable for use or to exclude it from further analyses.

In order to relate the results of assessment of correlations between the variables and to arrange the plants to the needs of a mining company through an expert analysis conducted at stage three, the author examined the set of representative variables for their influence on the value of an energy industry company's market attractiveness. A questionnaire, described in detail in [10], sent by post was used for this purpose. As the result of the expert analysis weights of the variables responsible for market attractiveness are obtained.

They are rated as follows:
- A variable indicated as very important by > 80% of respondents is assigned weight 1 (a very important parameter).
- A variable indicated as very important by 60–80% of respondents is assigned weight 0.75 (an important parameter).
- A variable indicated as very important by 40–60% of respondents is assigned weight 0.50 (a moderately important parameter).
- A variable indicated as very important by 20–40% of respondents is assigned weight 0.25 (a parameter of small importance).
- A variable indicated as very important by < 20% of respondents is assigned weight 0 (a parameter of no importance).

The proposed weighting system was already used by the author in the past in other research [13], [14].

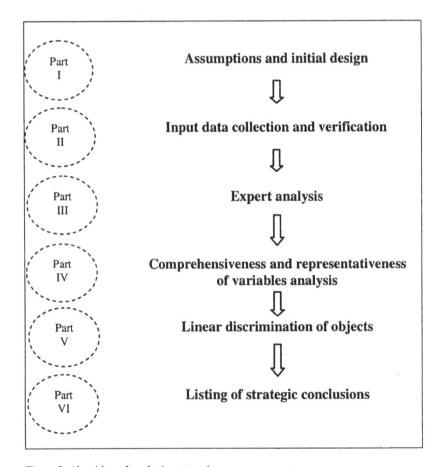

Figure 2. Algorithm of marketing attractiveness assessment

Before conducting further analyses a conglomerate weight of each parameter resulting from the experts' opinions needed to be taken into account.

To calculate it a method based on the arithmetic weighted average method was used. In the first step the effect of each individual potential on a plant's market attractiveness was calculated from formula (1):

$$W_{jPsr} = \frac{1}{m} \sum_{k=1}^{m} W_{jPk} \qquad (1)$$

where: W_{jPsr} – the weight of the j-th potential on the level of customer's market attractiveness, W_{jPk} – the weight of the j-th potential on the level of customer's market attractiveness assigned by the k-th expert, m – number of experts participating in the survey.

In the next step the effect of each individual potential on a customer's market attractiveness was calculated taking also into account values of W_{Psr}.

In order to calculate these parameters, average expert weights had to be calculated for all variables from formula (2):

$$w_{isr} = \frac{1}{m} \sum_{k=1}^{m} w_i \qquad (2)$$

where: w_{isr} – average weight of the i-th variable on the level of customer's market attractiveness, w_i – weight of the i-th variable on the level of customer's market attractiveness assigned by the k-th expert, m – number of experts participating in the survey.

To calculate the final values of weights W_i (used in further steps of the algorithm) the following formula was used (3):

$$W_i = \frac{w_{isr}}{\sum_{i=1}^{n} w_{isr}} \cdot W_{jPsr} \qquad (3)$$

where: W_i – weight of influence the i-th variable exerts on a customer's market attractiveness, n – number of variables accounting for the j-th potential.

In the questionnaire the experts were also asked to assign variables to one of the three groups: stimulants, destimulants or nominants. Additionally, in case of nominants, the experts were asked to give the optimal, in their opinion, value of every such variable. The final division of variables to stimulants, destimulants and nominants was done according to the majority of the answers – a variable was classified as a stimulant, a destimulant or a nominant if classified as such by at least 50% + 1 of all respondents.

Linear discrimination of the examined plants, which was planned as the 5th step of the assessment (Fig. 2) has to be preceded by a thorough examination of comprehensiveness and representativeness of the variables characterizing them. The method selected for the purpose was a multifaceted correlation of variables analysis. This type of analysis provides for examining mutual partial correlations between the analysed variables. After "data collection and verification" this stage of the survey allows to eliminate some of the initial 77 variables and so simplify the whole task.

The analysis of mutual correlations was done for variables belonging to the same potential groups. A correlation matrix was found to be the easiest way to calculate partial correlation indexes. For every column of each correlation matrix a threshold value of the linear correlation index r, represented by symbol r*, was calculated.

The procedure can be conducted with the use of formula (4):

$$r^* = \min_i \max_j \left| r_{ij} \right| \left(i, j = 1,..., k; i \neq j \right) \qquad (4)$$

where: r^* – threshold value of the linear correlation index r_{ij}, r_{ij} – index of correlation between the i-th and the j-th variable.

In order to verify the validity of variables based on the calculated correlation indexes r_{ij} and r^*, an analytical algorithm proposed by the author was used.

The key element of the algorithm was the results of the expert analysis or, to be more precise, its part where weights W_i of all variables affecting market attractiveness were established. The proposed selection criterion was that only variables with the greatest weights, i.e. those with highest importance, were accepted in further analysis. Due to their importance they were called dominant variables.

Basing on linear correlation indexes between each dominant variable and the remaining variables contributing to the potential, the next step of the analysis is to establish those of them for which the index of linear correlation with the dominant variable is greater than the threshold, i.e. (5):

$$r \geq r^* \qquad (5)$$

It was proposed that those variables, due to their high correlation with the dominant variable, were disregarded in further analyses.

The tools, proposed to be used at the initial stages of the process, devoted to collecting, verifying and analysing comprehensiveness of variables as well as establishing the weighing system, mark the starting point for classification of power plants. The classification stage has to be preceded by standardization of variables. Moreover, all variables have to be converted to a uniform standard. This was done by converting destimulants and nominants to stimulants. In the case of destimulants they were multiplied by (-1). In the case of nominants, inverses of modules of differences between the actual and the desired values of parameters were used.

Variables prepared in the way described above were ready to be classified. This was based on a standardised value sums method. In line with the nomenclature used in this work the method can be described as a sum of standardised values method broadened to incorporate assessment of customer market attractiveness index (WAK).

The sum of standardised values method consists in summing products of standardised variable values and the weights assigned to them by the experts.

Assuming that all variables were converted to stimulants, the calculation is done according to formula (6):

$$WAK_j = \sum_{i=1}^{n} w_i \cdot z_{ij} \qquad (6)$$

where: WAK_j – market attractiveness index of the j-th customer, w_i – weight of the i-th variable, z_{ij} – standardized variable, n – number of variables.

Results of the analyses proposed above are used as a decision-making tool in strategic planning. The main task is to divide plants into groups (segments) depending on their attractiveness (WAK value).

The author proposed that three strategic segments of power generation industry companies were created:

– segment A customers (key) – "high" WAK values;
– segment B customers (important) – "medium" WAK values;
– segment C customers (standard) – "low" WAK values.

Segmentation would be based on differences between the maximum and minimum WAK values. Segmentation of customers may form a basis for mining companies to create a differentiated, motivational customer management system, as the author intends to show in his subsequent works.

4. AN ILLUSTRATIVE CALCULATION OF POWER GENERATION COMPANIES MARKET ATTRACTIVENESS INDEXES

With the use of the methodology described in point 3 the author set on a task of ranking all 29 companies, incorporating all power and heat-and-power generation plants, operating in Poland, according to values of their company attractiveness indexes WAK. Values of 77 variables were calculated for the years 2003–2005. Due to confidentiality of some data used for the calculations the author could not include the values in this paper.

A time span of three years was chosen by the author to enable long-term observations of changes of indexes for individual companies.

In line with the proposed methodology, variable distributions were verified for their coincidence with normal distribution. All values of λ were lower than the values of λ_α read from Kolmogorow-Smirnow tables and at the same time values of W were lower than those appearing in Shapiro-Wilk test tables. This proved that the assumption that values of the variables were spread according to normal distribution was correct. An analysis of rare observations allowed for first verification of the list of analytical variables used in further analyses. Five variables from the financial potential group were eliminated.

As already stated in this paper the main objective of the work was to develop a ranking of power and heat-and-power plants according to the characteristics specified as important by relevant (customer service) mine employees. The necessary expert analysis was conducted, according to the rules described in point 3, at Kompania Węglowa S.A. – the biggest coal producer in Poland. The study was conducted from July 2006 through February 2007 on a group of 32 respondents selected by the company management.

For each potential and every variable average weight was calculated based on ratings given by the respondents using formulae (1), (2) and (3). The very interesting results of the survey could not, unfortunately, be presented here due to the limited space. They will be published by the author in his subsequent works. A very important conclusion emerging from the survey was that of all parameters affecting market attractiveness of companies in the power generation industry the greatest weight, in the opinion of mine officials, was carried by their production and sales potential. As slightly less important was viewed the company's current financial situation and its ability to meet their obligations to fuel suppliers. A surprisingly low weight was given to environmental potential. This may indicate that mine officials do not pay much attention to pro-environmental development of power plants and do not consider co-operation on developing environmentally safe methods of burning of coal as an important part of their relations with the customers. Such approach is contrary to the general directions set by the EU for the fuel and energy industry and regulated by applicable EU directives.

Most of the variables (56 out of the total of 77) were classified by the experts as stimulants, including the dominant variables of the production, sales and financial groups. The second largest group of variables were destimulants, with 19 variables. As a destimulant was also classified the dominant variable of the environmental potential group represented by symbol x_{72}. Only two variables: x_7 i x_8 were classified by the majority (not as prominent as in the case of other groups, though) of experts as nominants.

In the next stage some of the variables were eliminated due to high correlations between each other. The first step was to single out variables, which were given the highest weights by the experts under the assumption that they were the parameters exerting the greatest effect on the value of a company's market attractiveness.

They were:
- x_6 – volume of coal purchased – in the production potential group;
- x_{11} – quantity of electric power produced – in the sales potential group;
- x_{28} – liquidity index – in the financial potential group;
- x_{72} – volume of SO_2 emissions – in the environmental potential group.

The above variables were called dominant. In the second step boundary correlation indexes were calculated and comparing correlation indexes between every variable and the dominant in each group, as described in point 3, allowed to eliminate some of the variables, and thus decrease their number from 72 to 50. The biggest number of variables remained in the financial potential group. This is a direct result of low correlations between various variables in this group.

The most important part of the analysis, from the marketing viewpoint, was linear discrimination of objects using WAK index. This was preceded by standardization of variables and conversion of destimulants and nominants to stimulants.

The discrimination was done using formula (6). Plants were analysed mostly in an operational perspective (for the years 2003, 2004 and 2005) but from a strategic viewpoint as well (for the period 2003–2005). The results of calculations for the strategic analysis are shown in Figure 3.

Looking at the presented results one becomes immediately aware of big difference between the highest and lowest WAK value. This proves the general perception of vast gaps between different companies of the power generation sector in Poland.

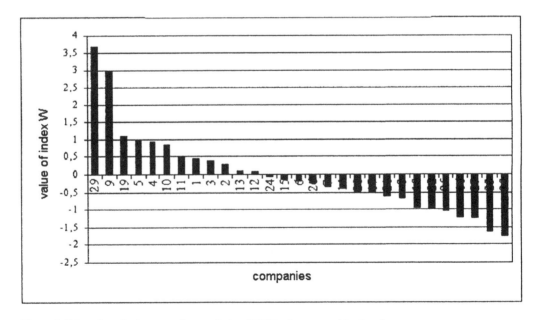

Figure 3. Value of marketing attractiveness index (WAK) of power and heat-and-power generating plants in Poland and a ranking of these companies in 2003–2005

Detailed analyses of WAK values for the years 2003, 2004 and 2007 showed that in the case of most customers similar value levels of individual indexes were observed. This was one of the reasons why companies virtually did not change their positions in rankings in consecutive years, with some of them keeping the same positions (companies with numbers 29, 9, 3, 12, 13, 15, 28 and 23).

This indicated that market attractiveness of these companies for a mining company remained stable for a prolonged period of time. When dealing with such companies this situation allows trade relations to be planned with low risk even if the plan is based on a short-term analysis.

Table 1. Segmentation of plants to strategic customer groups
in a customer relationship management system of a mining company

Ranking position	Plant number and its WAK value		Customer market attractiveness segment
1.	29.	3,6658043	Segment A (Key customers)
2.	9.	3,0090120	
3.	19.	1,1028222	Segment B (Important customers)
4.	5.	1,0020208	
5.	4.	0,9278023	
6.	10.	0,8477675	
7.	11.	0,5090824	
8.	1.	0,4484820	
9.	3.	0,4014789	
10.	2.	0,2939625	
11.	13.	0,1089158	
12.	12.	0,0824147	
13.	24.	−0,0511778	Segment C (Standard customers)
14.	15.	−0,1310822	
15.	6.	−0,1815834	
16.	26.	−0,2452559	
17.	7.	−0,3475199	
18.	14.	−0,3476219	
19.	20.	−0,5055930	
20.	17.	−0,5057308	
21.	21.	−0,6184363	
22.	8.	−0,6678756	
23.	18.	−0,9353206	
24.	22.	−0,9929407	
25.	25.	−1,0276622	
26.	16.	−1,2237308	
27.	27.	−1,2343999	
28.	23.	−1,6422052	
29.	28.	−1,7414294	

Exceptional to the described "stability rule" are several companies, which in the analysed period noted remarkable increases or decreases in values of their WAK indexes. The important fact is that more increases were noted among companies with "better" WAK values whereas more drops were among companies whose WAK indexes were already low.

Looking at the companies improving their ranking, attention should be paid to plant number 14, which climbed from position 23 in 2003 to 14 in 2005, and 19, climbing from 9 in 2003 to 3 in 2005.

The opposite can be observed for plant number 2, dropping from place 7 in 2003 to 12 in 2005, and 18, which in the same years ranked 22 and 25 respectively.

Beside the growth and decline trends spoken of above, worthy of attention is the phenomenon of sporadic fluctuations of WAK values observed in the case of a number of plants. The deviations observed there were not big but merit description. The fluctuations could be seen in the case of plants number 1, 10, 11, 24. The highest amplitude of fluctuations characterized unit 24, which ranked 10 in 2003, dropped to position 19 in 2004 and climbed back up to 10 in 2005. The observed changes of WAK were, in most cases, relatively high but short-term drops in 2004, which can be explained by a generally poor economic situation of those companies in that year.

In the last part of the algorithm, in line with the methodology described in Point 3, all companies were divided into three customer groups according to their market attractiveness. The results of the process are shown in Table 1.

CONCLUSIONS

The results of the analyses conducted by the author support a conclusion that the power generation industry, despite its apparent uniformity, is characterized by very high diversity. In view of the recently started liberalization of the Polish energy market, this finding is of utmost importance for coal producers constructing their marketing strategies. Doing away with long-term contracts in the situation when the global trend shows systematic increase in surplus of supply over demand will cause the coal market to become actually and truly free. Growing competition between coal producers makes them embark on marketing campaigns which, if no market analysis are done, may in many cases be inefficient. To avoid this, market monitoring systems should be used. The method presented in this work is, in the author's opinion, absolutely suitable for this purpose. It is compliant with the relationship marketing methodology, uses multidimensional comparison analysis tools and an expert analysis, and hence enables a complex analysis of the power generation industry to be performed, including segmentation of the market with respect to market attractiveness (WAK index values) of power and heat-and-power generation plants. Detailed knowledge of potential customers will allow mining companies to develop appropriate marketing strategies, using different marketing tools depending on the recipients' WAK value. The illustrative analysis shown in the work clearly reveals large differences between various companies in the energy sector as far as parameters, which are of vital importance to their supplier, are concerned. It shows also that the sector is made up of at least three different customer groups. The first is a stable group of leading companies with very high production efficiencies. The second group comprises customers characterized by medium and very unstable values of market attractiveness. Those customers, quite often equipped with high production potential, form a high-opportunity and high-risk segment. The last group are plants with low WAK index values. Those are highest-risk customers and use of marketing tools aimed at them should be limited.

REFERENCES

[1] Bogacz P. 2005: Rola marketingu w tworzeniu przewagi konkurencyjnej firmy górniczej w Polsce. Współczesne czynniki rozwoju przedsiębiorstwa, wyd. AGH, s. 271–284.

[2] Cheverton P. 2001: Zarządzanie kluczowymi klientami. Jak uzyskać status głównego dostawcy. Oficyna Ekonomiczna 2001.

[3] Mazurek-Łopacińska K. 2001: Orientacja na klienta w przedsiębiorstwie. PWE 2001.

[4] Chodorowska N., Krokosz E. 2002: Budowanie relacji z kluczowymi klientami. Marketing w praktyce, Informator, nr 1, 2.

[5] Borden N. 1964: The Concept of the Marketing Mix. Journal of Advertising Research, Vol. 4, pp. 2–7.

[6] Gronroos F. 1984: Idea of Relationship Marketing. Strategic Management. Concepts and Applications. European Journal of Operational Research, No. 26, pp. 23–47.

[7] Anton J. 1996: Customer Relationship Management. Prentice-Hall 1996.

[8] Morden T. 1991: Elements of Marketing. Prentice Hall 1991.

[9] [9] Lenskold J.D. 2003: Marketing ROI. The Path to Campaign, Customer and Corporate Profitability, McGraw Hill 2003.

[10] Bogacz P. 2007: Metoda oceny atrakcyjności rynkowej przedsiębiorstw energetyki zawodowej dla potrzeb budowy strategii marketingowej wielozakładowego przedsiębiorstwa górniczego. Praca doktorska, AGH.

[11] www.bip.gov.pl

[12] Draper N.R., Smith H. 1973: Analiza regresji stosowana. PWN 1973.

[13] Bogacz P. 2002: Program badania poziomu satysfakcji klientów jako obligatoryjne narzędzie platformy BTL dla producentów i importerów branży artykułów biurowych i szkolnych. Materiały Międzynarodowej Konferencji Studenckich Kół Naukowych: Nauki Ekonomiczne, s. 44–50.

[14] Bogacz P. 2004: Program monitorowania satysfakcji klientów kopalń – fundament marketingu relacyjnego w branży górniczej w Polsce. Zagadnienia interdyscyplinarne w górnictwie i geologii. Oficyna Wydawnicza Politechniki Wrocławskiej, nr 107, s. 69–81.

International Mining Forum 2008, Sobczyk & Kicki (eds) © 2008 Taylor & Francis Group, London, ISBN 978-0-415-46126-9

Conditional and Monte Carlo Simulation – the Tools for Risk Identification in Mining Projects

Leszek Jurdziak, Justyna Wiktorowicz

Institute of Mining Engineering, Wroclaw University of Technology, Poland

ABSTRACT: Based on the simple example of several investment schemes having the same expected value of cash flows but different volatility it has been shown that Monte Carlo simulation, when applied without any risk reduction techniques as e.g. hedging, does not reduce risk but only reveals it. Observation of cash flow variability in an investment scheme allows on proper selection of risk-adjusted discount rate for the given project. Monte Carlo simulation only shows potential risk, which can be measured also as the probability of loss, which increases with growing cash flow volatility despite expected NPV is constant. The risk of loss can be treated as the additional measure of risk connected with different investments. Introduction and popularity of VAR techniques in a finance sector encourages its application also in evaluation of mining projects. It has been also shown that the evaluation of mining projects should be based on the optimal life of mine schedules utilizing results of conditional simulation of ore body parameters in 3D due to only this approach allows on the proper estimation of cash flow variability in different years of mining project development.

1. INTRODUCTION

1.1. *Some confusion over the selection of discount rates in Monte Carlo simulations*

In the '90s in the Mining Engineering magazine, there was a discussion between Davis (1995) and Cavender (1992) due to confusion over the selection of discount rates (DRs) when performing Monte Carlo simulations (MCS) of a project's net present value (NPV) using risk adjusted discounting techniques. Cavender argued that lower DR should be used when performing stochastic valuation simulations, what creates higher NPV values and improves the attractiveness of a project. This is because the incremental uncertainty of a project has been recognized, in Cavender's view, directly in the spreadsheet cash flows of the simulation analysis. Therefore, it is no longer necessary to adjust project's DR upwards for its additional riskiness, even in risk-adjusted discounting paradigm. Davis argued that the DR should not be lowered. In his paper (1995), he demonstrated that stochastic simulations do not remove the risk inherent in a project, therefore, do not improve a project's value to the risk-averse investor. Risk adjusted discount rate (RADR) should, therefore, not be adjusted downward when performing stochastic NPV simulations.

The explanation and argumentation given by Davis was clear and should end the discussion however the problem of risk-free discount rate (RFDR) in MCS still comes back brought by different authors in different contexts (e.g. [15], [13], [16]). Attention is given by these authors to the individual investor's interpretation of the project's NPV distribution function (from MCS) based on its subjective preferences and risk tolerance in order to establish the real value of the mining project by certainty equivalent factors.

Unfortunately this approach depreciate a bit the value of MCS as the useful tool for assessing the real value of complex investment project and identification of risk connected with it by introducing a lot of subjectivity in interpretation of their results.

1.2. *Variability of cash flows (measure of risk) the base for RADR selection in MCS*

Base on the example of MCS of simple investments it will be shown that the attention should be focused not on the NPV probability distribution function (PDF), due to it is calculated with the questioned DR, but on the variability of projected (by MCS) cash flows (CFs) due to exactly this reveals risk connected with particular project. Based on the scale of this variability the risk-adjusted discount rate (RADR) should be chosen according to the riskiness of the project in comparison to other project realized by the analyzed company. The RADR should be calculated by addition of risk premium to weighted average cost of capital (WACC). Of course interpretation of the scale of variability and suited risk premium are subjective but the attained by MCS PDFs of CFs (their histograms) are fully objective if based on properly prepared input data for the MCS.

1.3. *The risk of loss*

MCS only shows potential risk, which can be measured also as the probability of loss, which increases with the growing cash flow volatility despite expected value of NPV is constant. The risk of loss can be treated as the additional measure of risk connected with different investments. Introduction and popularity of VaR and CFaR techniques in a finance sector encourages its application also in evaluation of mining projects.

1.4. *Usage of conditional simulation and pit optimization in mining risk identification*

It will be further shown that conditional simulation based on geostatistics and open pit optimization programs (NPV Scheduler v. 4) are the objective tools, which can be used for proper preparation of input data to MCS in order to attain annual CFs from optimal long term mining schedules.

The evaluation of mining projects should be based on the optimal life of mine schedule utilizing results of conditional simulation of ore body parameters in 3D due to only this approach allows on the proper estimation of CFs flow variability in different years of mining project development.

2. MCS OF THE EXAMPLE INVESTMENT SCHEME

2.1. *Assumptions*

The analyzed investment scheme is very simple. The net investment (NINV) in year 0 amounts 100 monetary units (m.u.; 1 m.u. = 10^6 PLN). Within the next 5 years the investors can expect positive annual CFs (30 m.u.). Assumed RFDR is 5% (e.g. interest rate of government bonds).

The safe investment (purchase of 5-years government bonds with guaranteed by government annual CFs for the next 5 years) has the definite NPV value 29.88 m.u..

There will be also considered several other investment schemes with volatile annual CFs (Fig. 1), what is a common situation in business environment. It is assumed that annual CFs will have the normal PDF with stable mean, which equals 30 m.u. and standard deviation from the range 1–60 m.u. – N(30, σ = 1–60) (Fig. 2). We can treat annual CFs as uncertain outcome from mining activity in different years. For simplicity it is assumed that volatility in each year is the same. In real life each CF within the whole life of mine (and even longer in order to include restoration CFs) have to be assessed individually based on all available data and different techniques (e.g. conditional simulation, pit optimization) used for their prognosis.

The outcome from analysis of such investments is presented in the next chapter.

2.2. Outcome of MCS

Several simulations (10 000 runs) have been performed for the assumed data in the Crystal Ball program (Fig. 2). Different analyses have been conducted showing more or less obvious results. It is self-evident that the expected NPV of all investments schemes will be the same (29.88 m.u.) due to symmetrical volatility of CFs having the same PDF in all years (Fig. 2).

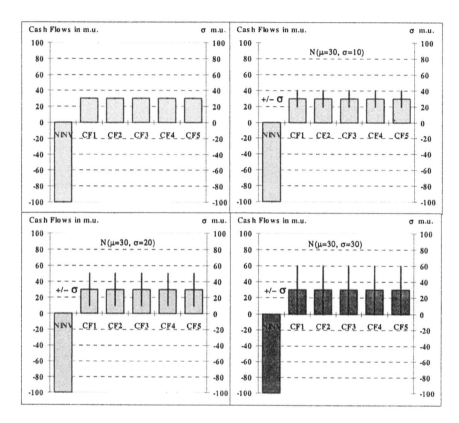

Figure 1. Example investment schemes having different volatility of CFs – N(30, σ)

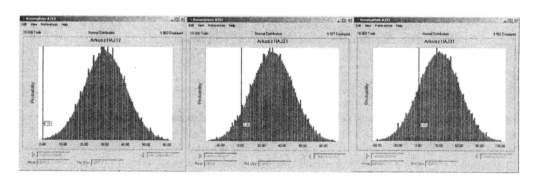

Figure 2. PDFs of CFs and their histograms from MCs for 3 inv. schemes: N(30, 10), N(30, 20), N(30, 30)

63

However with the increase of DR the NPV value will decrease showing the value of IRR in the intersection of the NPV line with the X axis (Fig. 3). For the safe investment there is no need to adjust DR upwards from 5% but for schemes with volatile CFs usage of RADR is self-imposed. Risk is commonly identified with volatility and standard deviation is used as the risk measure [8]. It can be seen that with the increase of CFs volatility more simulated CFs have negative value. In the first PDF (from the left side, Fig. 2) almost all CFs are positive while moving to the right the number of negative is increasing. In the third one (from the right side, Fig. 2) above 20% are below zero. Idea of CF at risk (CFaR) will be introduced later as a good measure of risk of investments based on probability of losses.

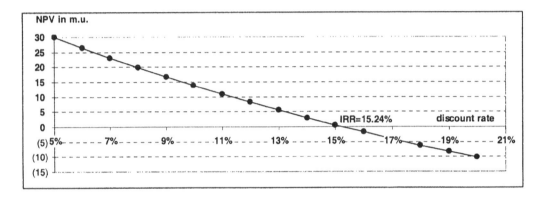

Figure 3. NPV as the function of discount rate (DR) for the safe investment scheme

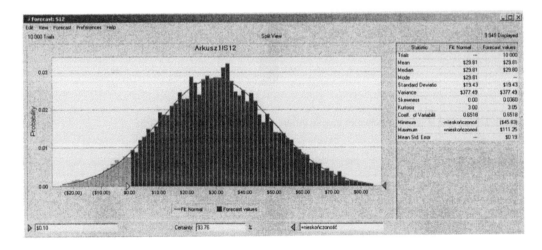

Figure 4. Histogram of NPVs for the CFs with the normal PDF N(30, 10) and DR = RFDR

Volatility of CFs is during MCS carried over to NPVs and exposed as their histogram (Fig. 4). Though before any NPV calculations the DR should be chosen. In the NPV with RADR method of investment profitability evaluation the RADR is established intuitively based on the decision ma-

ker's experience. In MCS the decision maker has an opportunity to observe individual CFs volatility (histograms of CFs) in each investment period before taking the decision, which RADR apply in
NPV calculations. In comparison to volatility of CFs in other projects realized by a particular company it is easy to find out if the DR should be on the WACC level or higher due to greater volatility.
Depending on observations of each CFs' variability even individual RADRs can be applied for each
CF in NPV calculations. Based on the general or individual RADRs the histogram of NPVs can be
created based on the MCS outcome. It means that the MCS should be carried twice. For the first time to select the proper RADR based on observations of resulting CFs – the exposed risk associated
with the analyzed project. For the second time to get the histogram of NPVs and carry out analysis
of MCS' results including sensitivity analysis and other specific statistical investigations. Of course
it can be done in one go by the calculation of NPVs for several DRs and taking into account only
this, which corresponds, to the chosen RADR.

3. INTRODUCTION OF THE NEW MEASURE OF RISK OF MINING PROJECTS

3.1. *New approach to risk and the "Dudek" Cabaret*

Very interesting aspect of investment risk can be described by the famous dialog between Bieniek
Rappaport (acted by Edward Dziewoński) and Kuba Goldberg (Wiesław Michnikowski) in the skit
"Sęk" ("Knag") performed many times in the "Dudek" Cabaret in Warsaw and played as the TV-
Show [7].

Bieniek is calling to Kuba with the proposal to finance a profitable investment:

Bieniek: There is a deal to make.

Kuba: A deal? How much I can lose?

Bieniek: Why you have asked me how much you can lose? You should ask me how much you
can earn.

Kuba: What can be earned, will be earned. I ask you: how much should I have to risk it, in case
we lose.

In this conversation one of potential investors (Kuba) focuses attention of the second one (Bieniek) to the possibility of losing investment capital, not on making profit what is the aim of most of
investors. Exactly this attitude to risk is embedded in the new method of risk management called
"Value at Risk" which played a key role in the development of financial risk management in the
banking sector in the last decade due to modifications of its supervision [17].

3.2. *Value-at-Risk and Cash Flow at Risk*

Value-at-Risk (VaR) measures the worst expected loss under normal market conditions over a specific time interval at a given confidence level. VaR answers the question: how much can I lose with
x% probability over a pre-set horizon. A VaR statistic is made up of three components: a time period, a confidence level and a loss amount (loss percentage).

For the given confidence level $\alpha \in (0, 1)$ the VaR of the portfolio at the confidence level α is given by the smallest number of l such that the probability that the loss L exceeds l is not larger than
$(1 - \alpha)$.

$$VaR^{\alpha} = \inf\{l \in R : P(L > l) \leq 1 - \alpha\} = \inf\{l \in R : F_L(l) \geq \alpha\} \qquad (1)$$

In probabilistic terms VaR^{α} is a quantile of the loss distribution F_L [12].

There are three ways to estimate Value-at-Risk: Historic simulations, variance-covariance calculations and MCSs [1].

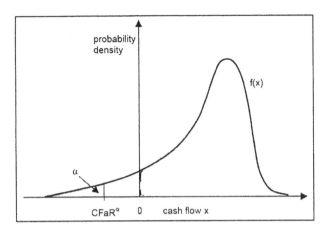

Figure 5. Cash Flow at Risk for probability level α (CFaR$^{\alpha}$):
$\alpha = P(x < CFaR^{\alpha})$ [17]

VaR is used mostly by companies in the finance sector but it can be also applied for a particular investment project. In such case we can use also Cash Flow at Risk (CFaR) which is the flow equivalent of VaR. While VaR is the worst loss over a target horizon with a given level of confidence on a stock represented by a financial asset the CFaR is the worst loss, out-flow of cash, over a given accounting period that an industrial investment project can cause to a non-financial firm (Fig. 5). The determination of a reliable value of the risk measure is important for its use. It is important for both the corporate governance (application in the principal-agent model) as well as the proper management of risk assessment in a particular company and its investments. Usage of MCS and VaR or CFaR techniques can be a good addition to traditional statistics based on histograms of CFs and NPVs such as: expected value and standard deviation.

4. PROBABILITY OF LOSS IN EXAMPLE INVESTMENT SCHEMES

Cumulative PDFs of NPV at DR = RFDR for different variability of CFs (Fig. 6) shows that the risk connected with investments schemes is growing with the increase σ. Although all investment schemes have the same expected value of NPV (E(NPV) = 29.88 m.u.) they can't be treated as the equal opportunities to earn money. Probability of loss or rather probability of not achieving safe earnings increases from 0 ($\sigma = 0$) up to almost 40% ($\sigma = 60$ m.u.) (Figs. 6 and 7a).

Because the aim of the company is the increase of its value through realization of profitable investments achieving the safe profit is not enough. Let assume that the company has the WACC = 10%, so 10% can be treated as the lowest rate of return the company should achieve from its investments – otherwise invested money will costs more than value of brought earnings. For the 10% DR at $\sigma = 10$ m.u the probability of not achieving required earnings (loss) is over 20% (Fig.7b). If the CFs volatility in other usually realized projects measured by coefficient of variation (CV) is only 0.1, what means that $\sigma = 10$% of mean (in our case $\sigma = 3$ m.u.) instead of WACC the RADR should be used – greater than 10% e.g. 13%. Risk of not achieving such return is over 35%. If the company has the opportunity to invest money to several alternative projects with IRR = 15% to compare the considered investment with them the opportunity DR should be used (instead of RADR). The NPV at DR = 15% will be positive (due to IRR = 15.24%, Fig. 3) but the risk of not achieving this result is close 50% (Fig.7b).

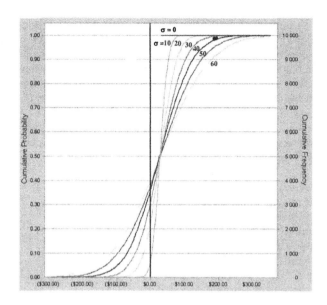

Figure 6. Cumulative NPVs' PDF for the CFs with different σ of CFs (0–60 m.u.)

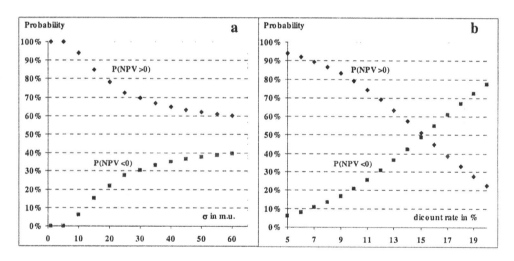

Figure 7. Probability that NPV is greater and lower than 0
as the function of σ at DR = RFDR (a) and DR at σ = 10 m.u. (b)

5. CS AND PIT OPTIMIZATION IN MINING RISK IDENTIFICATION

5.1. *Conditional simulation as the risk management tool in mining industry*

Risk Management has always been of prime importance for any mining company. The senior management spends a lot of money and resource to gain the confidence of the results of the deposit. The most sophisticated method to interpolate grade and classify reserve, till very recent, was that of the geostatistical methods. The latest trend, which supplements these methods and covers the risk of results, is Conditional Simulation (CS). It is a technique used to assess risk by means of a spatial Monte Carlo analysis. It provides a powerful tool for minimizing exposure of mining projects to risk.

CS reproduces the variability of the properties within an ore body and quantifies their variability; thus providing a realistic measure of the uncertainty or risk. Conventional point or block estimates such as those produced by kriging or inverse power of distance produce average values for properties, but the potential errors, both high and low, are functions of the variability. This gives CS a critical niche in the decision making process.

CS uses geostatistical parameters to provide several grade models, each of which are equally likely, honour the geological boundary constraints, honour the input data at sample locations and obey both the sample histogram and sample semivariogram [14].

Therefore CS gives the ability to:
– calculate confidence limits for grade estimates;
– calculate reserves above cut-off, without having to use support correction methods;
– assess the effect of different Selective Mining Units or bench heights;
– investigate the effect of changing sampling density;
– classify resources using a standardized repeatable procedure;
– determine the daily/weekly variation in mill head grade;
– assess the effect of stockpile blending;
– assess the sensitivity of the mine plan to the likely variation in grade estimates;
– optimize ore and waste block selection for grade control.

5.2. *The Datamine Studio Solution*

Although CS as the subject is 30 years old, it has taken a long time to become available for general use rather than just for professional geostatisticians.

The main reasons for it have been:
– the availability of suitable, easy to run software that doesn't require a degree in computing and geostatistics to run;
– suitable software which creates a good environment for processing and visualizing results, enabling their interpretation and further processing and analyzing in optimization and scheduling programs;
– speed of computers and software.

The newest Datamine Studio 3 software with Mineable Reserves Optimizer (MRO) module and NPV Scheduler v.4 pit optimization and scheduling program are suitable solutions of Datamine International Inc. to previously mentioned obstacles. The processes and CS approach used in Datamine Studio software san be seen on the flow diagram (Fig. 8).

The Datamine SGSIM simulation process is based on the most widely used algorithms for Sequential Gaussian Simulation described by Deutsch and Journel [6]. Two other GSLIB routines for transforming to and from Gaussian distributions have also been included.

A simulation study involves creating many equi-probable realizations, which are conditioned to the local data, and then analyzing the variability of the results. Datamine Studio provides a powerful set of tools for interpreting the results of the simulation. In addition to the standard data transformation and manipulation commands, two new processes have been added which deal specifically with the results of the simulation. One provides a detailed statistical analysis of the conditional distributions of the simulated model cells allowing confidence limits to be assigned (CSMODEL), and the other optimizes ore and waste block selection in a model by minimizing the loss due to misclassification (CSOWOPT).

The CS module combined with the existing statistical, geostatistical and data manipulation processes in Datamine Studio provide an environment for any conditional simulation study. Additionally the output can be easily fed into the Mineable Reserves Optimizer or NPV Scheduler for further investigation. The latest is especially important for lignite surface mining.

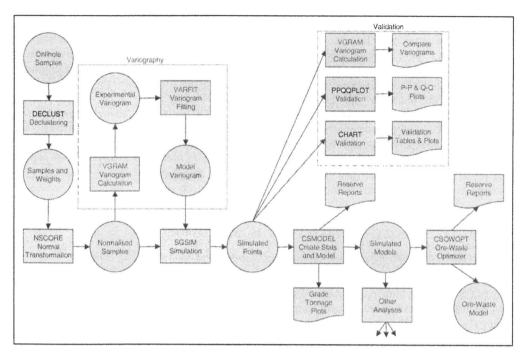

Figure 8. Approach to CS with procedures used in Datamine software [3]

5.3. *NPV Scheduler Methodology*

NPV Scheduler uses geological block models, economic models (based on mining costs, commodity prices etc.) and pit slope parameters to create Lerchs-Grossman nested pits (phases) within an ultimate pit. Then the optimal extraction sequence of blocks from each phase-bench is created that maximizes NPV using an algorithm that ensures contiguous mining progress on each bench.

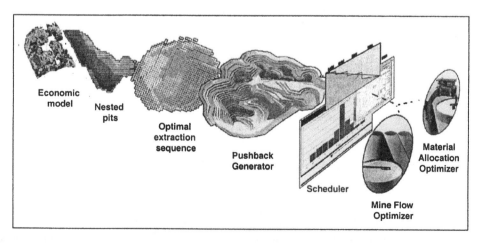

Figure 9. The sequence of data processing stages in Datamine NPV Scheduler v.4

The LG phases and the extraction sequence are then used to form pushback shapes according to economic value, production targets and engineering constraints that are practical to mine but produce maximum NPV. The pushbacks are then scheduled into a period by period production plan by forming "activities" out of mineable groups of blocks and using a powerful optimising engine to achieve maximum NPV while ensuring a steady flow of ore tonnes at mineable strip ratios and within the production constraints (Fig. 9, Datamine leaflet [4]).

NPV Scheduler v.4 offers Geo-Risk Assessment (GRA). GRA manages the uncertainty inherent in interpolating quality parameter (grade) distribution by considering conditionally simulated block models in the strategic planning process. GRA generates an ultimate pit for each conditionally simulated block model and calculates its NPV and profit. From the full set of ultimate pits GRA generates a range of risk-rated pits which become the basis for the strategic plan – limiting the impact of grade uncertainty on planning outcomes.

6. THE VALUE OF INTEGRATED STOCHASTIC SIMULATIONS

6.1. *Estimation of the LOM plan value based on CS, pit optimization and MCS results*

The evaluation of mining projects should be based on the optimal life of mine schedule utilizing results of CS of 3D ore body parameters and their Geo-Risk Assessment due to only this approach allows on the proper estimation of cash flow variability from pushbacks excavation in different periods of mining project development (Fig. 10).

Determination of all required for CFs calculation parameters (their histograms) can be done through CS, GRA and MCS based on all available data about the deposit, the mine and its economic environment at the moment of the schedule preparation. Therefore it has to be done in integrated environment allowing on analysis of several scenarios and case studies. The solution offered by Datamien software seems to be an ideal environment for such analysis.

Valuation of the investment project can be done inside NPVScheduler or outside it based on calculated parameters, which can be exported to the Excel spreadsheet. The optimal schedule can be prepared to maximize NPV or to maximize utilization of the deposit through its homogenization and usage of stockpiles (usage of MFO and MAO modules).

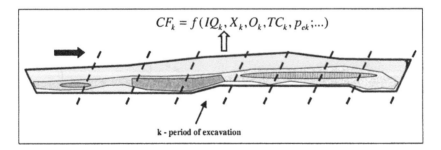

$$CF_k = f(IQ_k, X_k, O_k, TC_k, p_{ek}; ...)$$

k - period of excavation

Figure 10. The optimal Life of Mine schedule of lignite deposit excavation with CFs from pushbacks treated as random variables – functions of several conditional and MC simulated parameters

For each period (pushback) several parameters (their histograms) can be established in order to calculate CFs (2). These include such random variables as: quality indicator of lignite inside particular pushback (QI_k), amount of lignite (X_k), amount of overburden (O_k), total costs of mining and restoration (TC_k), price of electric energy (p_{ek}), etc. In calculation of CFs the outcome from econo-

70

mic analysis of operation of a mine and a power plant with usage of bilateral monopoly model, pit optimization and game theory should be used [9].

$$CF_k = f(QI_k, X_k, O_k, TC_k, p_{ek}; ...) \tag{2}$$

On the basis of calculated CFs the NPV of a project of lignite deposit excavation (or any other surface deposit) can be calculated (simulated) based on traditional formula but with random parameters (3).

$$NPV = \sum_{k=1}^{N} \frac{CF_k(...)}{(1 + RADR_{(k)})^k} - NINV \tag{3}$$

As the DR the RADR should be used due to observation of CFs variability should allow on the proper selection (even individual for each period k) of risk premium over WACC.

6.2. *The Foresight Project*

All previously mentioned Datamine software is available at the Institute of Mining Engineering at Wroclaw University of Technology for educational and consultancy purposes. Actually software for CS together with Crystal Ball program for MCS and real option analysis are going to be used at the Foresight Project "Scenarios of the Technological Development for the Lignite Mining and Process Industry" for risk analysis in the evaluation of planned excavation of the "Legnica" lignite deposit. Some results of MC simulations, but without CS application yet, have been already presented at the Polish Mining Congress [10]. Next results of risk analysis of a mine and a power plant bilateral monopoly [11] and results of CS and MCS should be published soon.

REFERENCES

[1] Benninga S. & Wiener Z. 1998: Value-at-Risk. Mathematica in Education and Research, Vol. 7, No. 4.
[2] Cavender B. 1992: Determination of the Optimum Lifetime of a Mining Project Using Discounted Cash Flow and Option Pricing Techniques. Mining Engineering, October 1992.
[3] Datamine Conditional Simulation in Datamine Studio, 2004 (slide presentation).
[4] Datamine NPV Scheduler v.4 leaflet. www.datamine.co.uk
[5] Davis G.A. 1995: (Mis)Use of Monte Carlo Simulations in NPV Analysis. Mining Engineering, January 1995.
[6] Deutsch C.V. & Journel A.G. 1998: GSLIB. Geostatistical Software Library and User's Guide, 2nd Ed., Oxford, New York, Oxford University Press.
[7] Dziewoński R. 1999: Sęk z Dudkiem. Wydawnictwo "Prószyński i S-ka", Warszawa 1999.
[8] Jajuga K. & Jajuga T. 2007: Inwestycje. Instrumenty finansowe, aktywa niefinansowe, ryzyko finansowe, inżynieria finansowa. Wydawnictwo Naukowe PWN.
[9] Jurdziak L. 2008: Analiza ekonomiczna funkcjonowania kopalni węgla brunatnego i elektrowni z wykorzystaniem modelu bilateralnego monopolu, metod optymalizacji kopalń odkrywkowych i teorii gier (Economic Analysis of Operation of a Mine and a Power Plant with Usage of a Bilateral Monopoly Model, Open Pit Optimization Methods and Game Theory). Oficyna Wydawnicza Pol.Wroc. (in press).
[10] Jurdziak L. & Wiktorowicz L. 2007: Elementy analizy ryzyka przy ocenie opłacalności produkcji energii elektrycznej z węgla brunatnego (Elements of Risk Analysis During Evaluation of Profitability of Energy Production from Lignite). Gospodarka Surowcami Mineralnymi, Kwartalnik, T. 23, Zesz. spec., Kraków 2007.
[11] Jurdziak L. & Wiktorowicz L. 2008: Identyfikacja czynników ryzyka w bilateralnym monopolu kopalni i elektrowni (Identification of Risk Factors in a Bilateral Monopoly of a Mine and a Power Plant). Górnictwo i Geologia X. Oficyna Wydawnicza Pol.Wroc. (in press).

[12] McNeil A., Frey R. & Embrechts P. 2005: Quantitative Risk Management: Concepts Techniques and Tools. Princeton University Press, Princeton 2005.

[13] Saługa P. 2005: Ocena górniczych projektów inwestycyjnych w aspekcie doboru stopy dyskontowej. Praca doktorska, Politechnika Wrocławska.

[14] Snowden D.V., Glacken I. & Noppe M. 2002: Dealing with Demands of Technical Variability and Uncertainty Along the Mine Value Chain. Brisbane, Qld, 7–8 October 2002.

[15] Torries T.F. 1998: Evaluating Mineral Projects, Applications and Misconceptions. Society of Mining, Metallurgy and Exploration Inc.

[16] Wanielista K., Saługa P., Kicki J., Dzieża J., Jarosz J., Miłkowski R., Sobczyk E.J. & Wirth H. 2002: Wycena wartości zasobów złoża. Nowa strategia i metody wyceny. Wyd. Instytutu GSMiE PAN, Biblioteka Szkoły Eksploatacji Podziemnej, Kraków 2002.

[17] Werkmeister C. 2005: Adjusting Financial Risk Measures to Industrial Risk Management. Challenges for Industrial Production. Workshop of the PepOn Project: Integrated Process Design for Inter-Enterprise Plan Layout Planning of Dynamic Mass Flow Networks, Karlsruhe, 7–8 November.

Publication sponsored by the Foresight Project "Scenarios of the Technological Development for the Lignite Mining and Process Industry" No. WKP_1/1.4.5/2/2006/4/7/585/2006 under Sectorial Operational Program – Increase of Competitiveness of Companies, 2004–2006.

International Mining Forum 2008, Sobczyk & Kicki (eds) © 2008 Taylor & Francis Group, London, ISBN 978-0-415-46126-9

Inherent Conflict of Individual and Group Rationality in Relations of a Lignite Mine and a Power Plant

Leszek Jurdziak

Institute of Mining Engineering, Wroclaw University of Technology, Wroclaw, Poland

ABSTRACT: Based on the newest results of economic analyses of operation of a lignite opencast mine and a power plant treated as a bilateral monopoly their relation has been described using game theory language. In such a tandem there is an inherent conflict of individual and group rationality – a conflict of maximization of own profit of both sides and maximization of joint profits. The mine has the information advantage (knows the deposit) and can always apply its predominant strategy – the optimal adjustment to changes of lignite prices. In order to secure acceptance of the solution maximizing joint profits (optimal in the Pareto sense) it is proposed to treat the negotiation between lignite mines and power plants as a cooperative, two-stage, two-person game with non-zero sum in which in the first stage both sides should select the ultimate pit maximizing joint profits and in the second one the agreement should be achieved regarding profit division (determination of the transfer price of lignite). It is proposed to use Nash bargaining solution in finding the accepted profit division but due to the conflict of interests only the full vertical integration of both sides can secure the Pareto optimal solution.

1. THE INFLUENCE OF LIGNITE OPENCAST MINE OPTIMISATION

Analysis of mine and power plant operation with usage of a bilateral monopoly (BM) model and pit optimisation [2], [3] has shown that:
- The tandem of an opencast lignite mine & a mine-mouth power plant as oppose to a classical BM has the determinate solution not only in the quantity of intermediate product (lignite) but also in its price.
- The optimal lignite price p_{bm} with corresponding optimal ultimate pit, which maximises joint profit of the whole BM in long run (its non-discounted value Π_{Vmax}) can be found for given economic conditions, costs structure, future demand for electric energy or its prices [2], [3].
- This optimal lignite price p_{bm} determines the division of profit between the mine Π_K and the power plant Π_E (Fig. 1), what in fact excludes price negotiation, unless:
  - an area around the maximum of joint profits is almost flat or there are few local maximums on the same level and it is possible to depict the new contract curve (Fig. 1); or
  - during negotiation, in conditions of cooperation and mutual trust, other split of joint profits has been decided and realized through:
    - side payments (in order to decrease the profit levels differences or realize agreed division);
    - determination of other lignite price – the transfer price p_{wt}, which will be used only for clearing accounts between both sides of BM in order to attain the agreed profit division.

In such a case both the selection of ultimate pit as well as other decisions about its shape and size change should be done based on optimal price (economically justifiable) maximizing joint profits of the whole system in order to keep economic rationality and effectiveness of mine activity. Usage of transfer prices other than optimal requires close cooperation, so it is rather possible in a holding or in a vertically integrated firm.

- Necessary condition for finding the optimal solution for BM is the realization of pit optimisation and parameterisation process. Sensitivity analysis of the size and shape of ultimate pit on changes of lignite base price is necessary to determine the influence of base price on long run supply of lignite, change of averaged quality parameters and costs of a power plant and a mine [12]. For this purpose it is also required to take into account the electric energy market through usage of long term forecasts of electric energy demand and expected level of electricity prices.
- Realization of open pit optimisation process is not possible without prior creation of 3D structural and quality models of the lignite deposit which are a base for the value model [1]. Geological modelling can be done in special mining and geology software such as Datamine Studio. The value model can be created there or in the pit optimisation software e.g. NPVScheduler+.
- Described procedure of selection of ultimate pit maximising joint profits of BM can be used for optimal, 3D delimitation of mineable reserves based on economic criteria connected both with a mine as well as a power plant and electric energy market on which both are active.

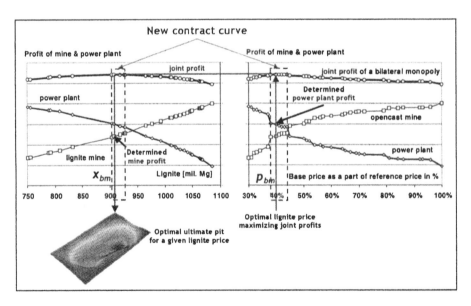

Figure 1. Joint profits of a mine and a power plant as a function of lignite quantity when the electric energy price $p_e = 0.13$ zł/kWh – the hypothetical solution for the pit placed on the "Szczerców" deposit [7]

2. NEGOTIATION BETWEEN POWER PLANT AND MINE AS A COOPERATIVE, TWO STAGE, TWO-PERSON, NON ZERO-SUM GAME

Utilization of the BM model and methods of open pit optimisation allows on treatment of lignite price negotiation between mine and power plant as a co-operative, two stage, two person, non zero-sum game [8]. Due to improvement in economic results should be found in optimal adjustments of shape and size of the ultimate pit to electric energy demand and not in the prolonged negotiation of lignite price the game should be carried in two stages.

In the first one (strictly cooperative) both sides on the basis of disclosed cost data and outcomes of the parameterisation process of lignite deposit should choose ultimate pit maximising joint profits of the whole system. The first stage would be the two person, non zero-sum game.

In the second stage, if the division of profit implicated by optimal solution is not satisfactory for one of sides, both of them should agree on different division and on connected with it lignite transfer price. So the problem of choosing appropriate transfer price would be only a technical one. It should be stressed that the agreed and accepted split of profit would determine the transfer price and not the other way round. This stage would be the positive-sum game in which the sum is equal to maximal profit Π_{Vmax}.

In the mine there would be two lignite prices: the economically justified (determined by the solution of modified BM of lignite mine and power plant) and the transfer price – used only for mutual clearing of accounts – realization of agreed profit division.

The organizational and ownership structure of BM of a mine and a power plant and attitude of both sides to mutual negotiation play a key role in its functioning and possibility of joint realization of the optimal solution – maximization of joint profits. There are inherent contradictions in BM such as: conflict of individual and group rationality and asymmetry of information, which could lead to solutions, which are not optimal in Pareto sense.

3. INHERENT CONFLICT OF INDYWIDUAL AND GROUP RATIONALITY

3.1. *Lignite prices in conditions of their control and confirmation*

The new "Energy Industry Law" voted in 1997 introduced the requirement of lignite prices confirmation by the President of the Energy Regulatory Office and precisely regulated formation of lignite prices. Base prices of referenced lignite could cover only justified expenses of mines (which were described in detail by this law), margin of profit not greater than 10% and appropriate tax. In paper [4] trials of mines to increase lignite prices during the period 1998–2003 and confirmed lignite prices for different mines has been presented (Fig. 2).

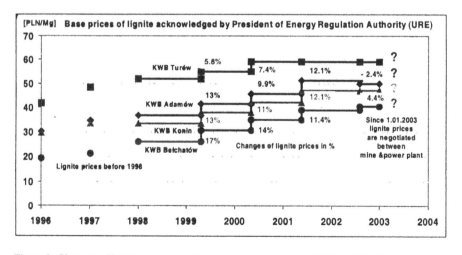

Figure 2. Changes of lignite prices in different mines in the period 1996–2003 [4]

On the 1st of January 2003 the "Energy Industry Law" has been changed and the confirmation requirement was cancelled, what eventually ended the period of control and confirmation of lignite prices in Poland. Since that moment lignite prices can be freely negotiated between power plants and mines.

This new situation has opened the area of research and analysis of BM operation and the role of lignite price in relation of both sides on a free market.

3.2. *Asymmetry of information and the predominant strategy of lignite mine*

In BM of a mine and a power plant this is the mine, which has the information advantage over the power plant. This advantage results from the knowledge of the lignite deposit. The mine knows the quality of lignite in the area of planned excavation and can forecast cost of its excavation based on data about the shape and size of the deposit as well as the amount of overburden and lignite. Of course this knowledge is only estimated due to the discrete identification of deposit features. Nevertheless the mine has this knowledge and the power plant not. This knowledge alone does not create any advantage especially in short term. At best it can be used as an excuse for difficulties with meeting agreed targets (regarding time, amount and quality of supplied lignite). This is why the optimised blending should be used for short term scheduling [13]. However in the long run the mine can for each lignite price find out the best ultimate pit maximising the net, non-discounted cash flows using Lerchs-Grossmann optimisation technique (Fig. 1).

In the alternative approach L-G ultimate pits can be replaced by the set of optimal schedules within optimal pits generated in optimisation software (e.g. NPVScheduler). Even though this approach is better because of usage of discounted values and schedules but it is much more complicated due to the scale of detail planning connected with generation of optimal excavation schedules for several lignite prices.

The possibility of the optimal adjustment of shape and size of the left for excavation part of the deposit creates the predominant strategy for the mine. The power plant decreasing the lignite price during negotiation never knows if the mine optimally adjusting to it does not resign from the excavation of the current pit in aid of excavating the smaller one but bringing more profit (due to its optimality) than the previous pit with smaller price. Due to the changes are not linear sometimes the small decrease of lignite price can lead to big difference in size and shape of the pit (Fig. 1). In consequence instead of expected increase of the power plant profit both the mine and the power plant can have lower profits in long run.

For each of nested pits it can be found two lignite prices: p_{iE} and p_{iK} (2) for which the mine and the power plant (respectively) attain break-even points (their revenues cover costs). There is also the lignite price p_i, which maximises net value of this pit Π_{iK} in comparison to profits offered by the rest of nested pits. Such price exists due to this pit is one of nested pits generated in parameterisation process and it is therefore the optimal pit for the given price p_i. This price, which can be called the border price, is the lowest lignite price for which this pit offers the highest profit Π_{iK}. If the next border price $p_{(i+1)}$ for the next pit (i + 1, a greater one) is lower then p_{iK} then for the prices from interval $[p_{(i+1)}, p_{iK}]$ it is better for the mine to excavate one of next pits. For prices from border prices interval $[p_i, p_{(i+1)}]$ profit of the mine is linearly increasing from Π_{iK} and the pit i is the best (gives the highest net cash flow from other pits). For the border price $p_{(i+1)}$ the net cash flows $\Pi_{(i+1)K}$ is greater and it is better for the mine to excavate the next pit (i + 1) than the previous one.

3.3. *Lignite price as a profit sharing tool in Nash bargaining solution*

In case of negotiation between mine and power plant the Pareto optimal set is the line of joint profit $\Pi_K + \Pi_E = \max(\Pi_V) = \Pi_{Vmax}$ (where $x = \Pi_K$ and $y = \Pi_E$) drawn for the ultimate pit (one of phases) which maximises joint profits for a given cost structure and future level of electricity prices (treated as a profit of vertically integrated energy producer Π_V). It is the line with slope coefficient -1 located furthest from the point (0,0) and going through one of points representing optimal phase (Figures 3, 4). Methodology of finding the optimal pit has been described earlier in papers [2], [3].

The area of grey triangle located in the 1st quarter of coordinate system (a positive one) below the Pareto optimal line (Fig. 4) can be treated as the acceptable set of solutions of negotiation/bargaining between mine and power plant. Nor the mine neither the power plant will voluntary accept loses when their joint profit is positive.

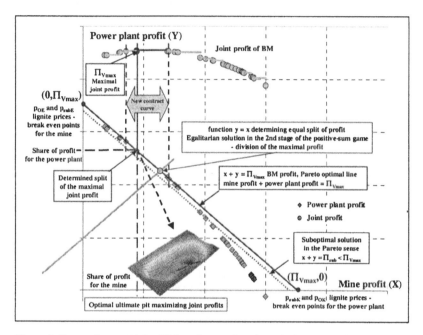

Figure 3. Power plant and joint BM profits as a function of mine profits for 35 ultimate pits generated for the "Szczercow" deposit for different lignite prices and electric energy price 0.13 zł/kWh [7]

It is impossible to attain greater profit than Π_{Vmax}, so all solutions should be in the triangle determined by points: (0, 0), (Π_{Vmax}, 0) and (0, Π_{Vmax}).

In case of linear transferability of utility function of payoffs (in this case a profit is treated as the payoff in bargaining game) the problem of profit sharing is much simpler and appropriate Nash solution is given by point (x_0, y_0) [14]:

$$x_o = 0.5 \ (x^* - y^* + \Pi_{Vmax}) \text{ and } y_o = 0.5 \ (y^* - x^* + \Pi_{Vmax}) \tag{1}$$

where: $\Pi_{Vmax} = x + y$ and (x^*, y^*) represents coordinates of the status quo[1] point.

It easy to see that $x_o + y_o = \Pi_{Vmax}$ and $x_o - y_o = x^* - y^*$, so in this case mutual location of both sides is kept unchanged and surplus utility of payments is equally divided between both sides [14].

[1]Status quo (Latin) currently existing state. Allocation of payoffs k* and e* for both sides (in advance established point in set of acceptable solutions), which follows if negotiation will prove abortive [21].

Graphical solution can be found on the intersection of line with slope coefficient +1 going through the status quo point and the negotiation set on the Pareto optimal line (N) (Fig. 4). The division of profit determined by the optimal lignite price p_{bm} (O) can be placed outside negotiation set (Fig. 4), what means that it can not be the solution of the game. The mine can not expect to get profit Π_{oK} due to the power plant profit Π_{oE} is below Π_E – its safety level[2] determined by the status quo point. It should be remembered that all data are only hypothetical and mutual location of all points in real situations can be different (compare with Fig. 3). Profits, the optimal lignite price, border prices of lignite determining break even points for mine and power plant as well as lignite price contours have to be calculated individually for the particular deposit based on revenues and costs of a mine and a power plant and taking into account their specificity (Fig. 5).

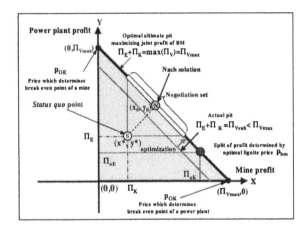

Figure 4. Set of acceptable solutions – game outcomes (the area of grey triangle), the status quo point (S) determining negotiation set on the Pareto optimal line plus the Nash solution (N) and the optimal solution (O) [7], [10]

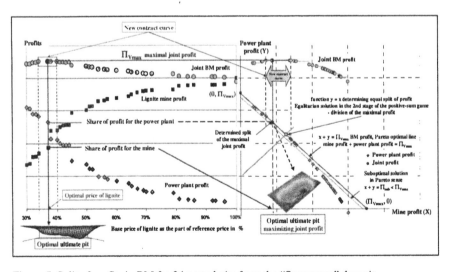

Figure 5. Split of profits in BM for 34 nested pits from the "Szczercow" deposit

---

[2]Safety level – a payoff in an optimal strategy of a player in a non-zero sum game.

3.4. *Lignite price contours*

The joint BM profits for the particular pit (being the sum of power plant and mine profits) is constant and does not depend of lignite price $(\Pi_{iK}(p) + \Pi_{iE}(p) = \Pi_{iV}(p) = const)$. The mine profit $\Pi_{iK}(p)$ for the lignite price p from the break-even prices interval (p_{iE}, p_{iK}) increases from 0 up to Π_{iV} and the power plant profit decreases from Π_{iV} down to 0. This means that on lines $x + y = \Pi_{iV} = const$ in the positive quarter of coordinate system we can place lignite prices starting from p_{iE} up to p_{iK} in intervals proportional to the quotient[3] $(p_{iK} - p_{iE})/SQRT(2\Pi_{iV}^2)$. It is enough therefore to find out break-even prices (2) for each of nested pits [2], [3].

$$p_{iE} = \frac{c_K(x_i)}{x_i}, p_{iK} = \frac{p_e e(x_i) - c_E(x_i)}{x_i} \tag{2}$$

where: $c_K(x_i)$ is the total cost of excavating x_i amount of lignite from the pit i, p_e is the expected future price of electric energy, $e(x_i)$ is the amount of electric energy produced from x_i amount of lignite, $c_E(x_i)$ is the total cost of energy production form supplied lignite (without cost of lignite purchase).

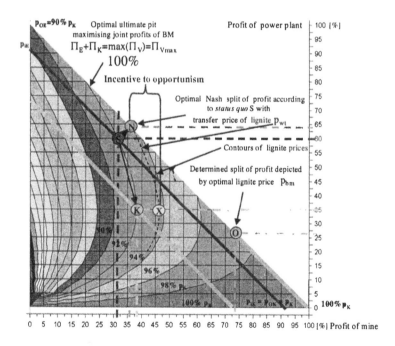

Figure 6. Usage of lignite price contours in profit division in BM and demonstration of the inherent contradiction of individual and group rationality. Lignite contours are drawn every 2% [10]

The maximal joint profit which can be attained in BM in given economic conditions from excavation of the optimal ultimate pit is equal $\Pi_{Vmax} = max\{\Pi_{iV}\}$. Profits from this and other pits and

---

[3]SQRT – square root.

shares of these profits falling to the power plant $\Pi_{iE}(p)$ and the mine $\Pi_{iK}(p)$ for different lignite prices p (Fig. 6) are expressed as the percent of maximal profits Π_{Vmax}. Similarly the lignite price is expressed as a percent of the break-even lignite price for the power plant p_{iK}. This price due to the simplified assumption regarding direct proportionality of the lignite productivity $e(x_i)$ and total power plant costs $c_E(x_i)$ from the amount of lignite x_i is stable $p_{iK} = const = p_K$. It has been also assumed that the break-even lignite price p_{iE} for the mine changes from 65% up to 90% of p_K –90% for the optimal ultimate pit, and 65% for the smallest pit. It is assumed that for each number from 0 up to Π_{Vmax} it can be found the optimal ultimate pit bringing exactly this profit. Such assumptions are arbitrary and can be far from the real situation. Usually number of nested pits is restricted (in the "Szczercow" case it was 34 pits) and the profit and break-even prices changes are not linear (Fig. 1). However it is used only to demonstrate the influence of pit selection on joint profits and the negotiated price on split of profit. It also illustrates growing incentive to opportunism with the increase of difference between the optimal and the negotiated price (Figs. 6–7).

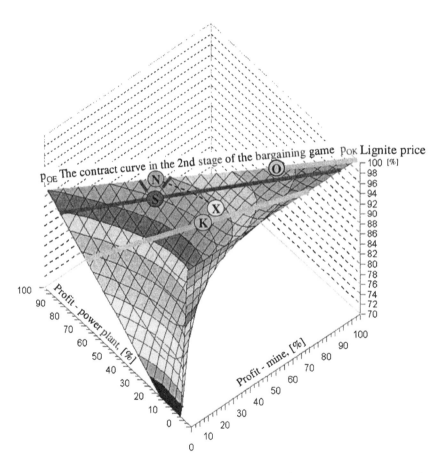

Figure 7. Acceptable solutions with hypothetical contours of lignite prices as a function of two profits, the status quo point (S) determining negotiation set with the Nash solution (N) and the optimal solution determined by the optimal price p_{bm} (O) plus the contract curve in the 2nd stage of the game – the Pareto optimal line between prices p_{OE} and p_{OK} [7], [10]

3.5. *Incentive to opportunism*

Suppose that the actually excavated pit is not optimal and gives only 91% of the maximal profit and the negotiated earlier price (92% p_K) determines status quo point S. In such situation the mine has two solutions. Preferring cooperation it can select the pit optimal for the BM (giving 100% of maximal profit $\Pi_{V\text{max}}$) and calculate the new transfer price for the Nash solution (point N), which will be a bit higher (about 1.8% p_K) what improves its share of profit from 31% up to 36%. The power plant share of profit also increases from 60% to 64% (the movement from point S to N, (Figs.6–7)). Alternatively, behaving opportunistically (if there is no cooperation), the mine can select the variant optimal only for itself (excavation of the smaller pit – the movement from point S to K.

The choice of the predominant strategy means the increase of its shares of the long-term profit from 31% up to 38%. Unfortunately for the power plant it means decrease of profit from 60% down to only 35%. The excavated pit would generate joint profits on the much lower level, which equals only 73% of the maximal profit (BM would loss 18% of potential profits). The mine opportunism could be even not noticed due to both the actual lignite price and contracted annual amount of lignite is not changed. Only the ultimate pit is different so the changes would be seen only in long run as the shorter period of pit excavation and lowered total profits, what can be explained by more difficult geological conditions than expected before.

The difference between profits attained by the mine selecting the Nash bargaining solution (point N) and choosing the predominant strategy (point X) creates the incentive to opportunistic behaviour (Fig. 6). This incentive decreases with the increase of the lignite price approaching the optimal lignite price (movement of point N in direction of point O). In the "Szczercow" case the egalitarian solution (equal split of profits) lies on the new contract curve – optimal and egalitarian points are close to each other (Figs. 3 and 5). It is therefore necessary to check for each deposit and each set of economic conditions how big is the inherent conflict of individual and group rationality in order to reduce it and prevent opportunistic behaviour through the appropriate transfer price selection [9].

Additional space is required for the discussion of proper choice of status quo point and its usage in tactical and strategic negotiations as well as treatment of described here optimisation process as a real option to change the scale of mine activity.

CONCLUSIONS

Organizational and ownership solution of mines and power plants and their attitude to negotiation have a key influence on BM operation and capability of realisation of optimal variant maximising joint profits. Analysis of different BM organizational and ownership structures [6] leads to conclusion that because of the inherent conflict of individual and group rationality[4] [5] only the full vertical integration of the lignite mine and the power plant can secure realisation of the optimal solution – excavation of the optimal ultimate pit maximising joint BM profits [2], [3].

In any other structure the incentive to opportunism can appear [6]. It magnifies with the increase of the difference between the optimal lignite price p_{bm} and the price established during negotiation p_{wt} (Fig. 6). It is therefore important for the power plant to know the optimal price and to have equal access to all necessary information.

Asymmetry of information also leads to opportunism. Knowledge about the deposit in condition of lack of cooperation and concentration only on price negotiations allows the mine to apply its predominant strategy – for every negotiated lignite price it can choose the ultimate pit, which maximises its own profits. Usually it will be the smaller pit than those optimal for the whole BM, what re-

---

[4]This contradiction can be seen on lignite price contours drawn on the profit division chart (Fig. 6) [9].

duces joint profits and profits of the power plant in long run. In consequence the period of excavation will be shorter and utilisation of the deposit lower. It means that this solution is not optimal in Pareto sense [8].

Two different owners of BM sides having opposed interests increase the threats of non-cooperative behaviours and realisation of the sub-optimal solution.

A profit of BM doesn't depend on lignite price. It means that its improvement should be found in optimal adjustment of the shape and size of the ultimate pit to demand for electricity (its prices) and not in prolonged negotiations of lignite prices. Lignite price decides only about the profit division between a power plant and a mine and cannot improve joint financial results.

It is proposed to use Nash bargaining solution to divide profit in a bilateral monopoly of a lignite mine and a power plant. As the Nash solution is the only bargaining solution, which is effective, symmetrical, scale covariant, and independent of irrelevant alternatives it is commonly accepted as the equitable solution in bargaining problems.

Application of Nash bargaining solution moves the difficulty from finding the solution (fair profit division) to proper selection of the status quo point. As the status quo it can be taken safety levels in a game or any other agreed or forced (e.g. by applying strategy of rational threats) profit division. It depends from bargaining parties what objective and rational arguments they will use in condition of co-operation or what threats they will announce and will be ready to apply in condition of rivalry and lack of cooperation.

Vertical integration of lignite mines and power plants does not create any threats to electric energy market and electricity consumers due to the pit optimal for BM is greater than the pit optimal only for the mine and the energy supply increases. There will be also other positive synergy effects for consumers (e.g. decrease of transactional costs) increasing economic effectiveness of the integrated energy producer. First of all only the full vertical integration of both sides can secure the Pareto optimal solution and eliminate the inherent conflict of individual and group rationality from relations of a lignite mine and a power plant [9].

More detail explanation of presented matters can be found in the prepared monograph ([11] in press).

REFERENCES

[1] Jurdziak L. 2000: Principles of Creation of 3D Value Model and Mining Cost Models for Optimisation Programs (in Polish). Gornictwo Odkrywkowe (Opencast Mining), No. 1, 2000.

[2] Jurdziak L. 2004a: Tandem. Lignite Opencast Mine & Power Plant as a Bilateral Monopoly. Mine Planning and Equipment Selection, A.A. Balkema Publishers, Taylor & Francis Group plc., 2004.

[3] Jurdziak L. 2004b: The Influence of Lignite Opencast Mine Optimisation on Solution of Bilateral Monopoly Model of Lignite Mine & Power Plant in Long Run (in Polish). Gornictwo Odkrywkowe (Opencast Mining), No. 7, 8, 2004.

[4] Jurdziak L. 2005a: Forming Lignite Prices in Conditions of Their Control and Confirmation (in Polish). Gornictwo Odkrywkowe (Opencast Mining), No, 4, 5, 2005.

[5] Jurdziak L. 2005b: Is Vertical Integration of Mines and Power Plants Profitable and for Whom? (in Polish). Biuletyn URE (Bulletin of the Energy Regulatory Office), No. 2 (40), 2005.

[6] Jurdziak L. 2006a: Influence of Structure and Ownership of Lignite Opencast Mine and Power Plant Bilateral Monopoly on Its Operation. Proceedings of the International Conference Mine Planning and Equipment Selection, Torino 2006.

[7] Jurdziak L. 2006b: Lignite Price and Split of Profit Negotiation in Bilateral Monopoly of Lignite Opencast Mine and Power Plant. Proceedings of the International Conference Mine Planning and Equipment Selection, Torino 2006.

[8] Jurdziak L. 2006c: Lignite Price Negotiation Between Opencast Mine and Power Plant as a Two-Stage, Two-Person, Cooperative, Non-Zero Sum Game. Proceedings of the International Symposium Continuous Surface Mining, Aachen 2006.

[9] Jurdziak L. 2007a: Lignite Prices as Transfer Price. Part One: Introduction, Part Two: Law Aspect and Part Three: Economics Aspect (in Polish). Przegląd Górniczy (Mining Review), No. 6, 7, 8, 2007.

[10] Jurdziak L. 2007b: Nash Bargaining Solution and the Split of Profit in Bilateral Monopoly of Lignite Opencast Mine and Power Plant. Part One: Theoretical Foundations and Part Two: Applications in Strategic and Tactical Negotiations (in Polish). Gornictwo Odkrywkowe (Opencast Mining), 2007.

[11] Jurdziak L. 2008: Economic Evaluation of a Lignite Mine and a Power Plant Operations with Osage of a Bilateral Monopoly Model. Pit Optimisation and Game Theory. Oficyna Wydawnicza PolWroc (in press).

[12] Jurdziak L., Kawalec W. 2004: Sensitivity Analysis of Lignite Ultimate Pit Size and Its Parameters on Change of Lignite Base Price (in Polish). Mining and Geology VII. Scientific Papers of the Institute of Mining Engineering of the Wroclaw University of Technology, No. 106, Studies and Research, No. 30, Wroclaw 2004.

[13] Kawalec W. 2004: Short-Term Scheduling and Blending in Lignite Open-Pit Mine with BWEs. Mine Planning and Equipment Selection, A.A. Balkema Publishers 2004.

[14] Owen G. 1975: Game Théory (in Polish). PWN, Warszawa 1975.

Publication sponsored by the Foresight Project "Scenarios of the Technological Development for the Lignite Mining and Process Industry" No. WKP_1/1.4.5/2/2006/4/7/585/2006 under Sectorial Operational Program – Increase of Competitiveness of Companies (2004–2006).

International Mining Forum 2008, Sobczyk & Kicki (eds) © 2008 Taylor & Francis Group, London, ISBN 978-0-415-46126-9

Method of Identification of Mineable Lignite Reserves in the Bilateral Monopoly of an Open Pit and a Power Plant

Leszek Jurdziak, Witold Kawalec
Institute of Mining Engineering, Wroclaw University of Technology, Poland

ABSTRACT: In comparison to old approach used in Poland for ore reserves and mineral resources classification the new methodology of geological, mining and economic data processing is proposed to estimate mineable reserves of lignite for electric power generation. In order to achieve Pareto optimal solution which assures the highest level of resources utilization it is necessary to apply the pit optimization methods and the newest results of the analysis of modified bilateral monopoly of opencast lignite/coal mine and mine-mouth power plant.

1. INTRODUCTION

The methods of ore reserves and mineral resources classification and reporting have been changed over the years, mostly independently in different countries. They still have not reached the internationally valid standard despite the fact that significant mining companies usually carry on their mining operations in many countries simultaneously. To overcome this problem some efforts have been undertaken to develop the standardised ore reserves classification [13] among which the most successful seems to be the Australasian Joint Ore Reserves Committee (JORC) standard. The JORC code, first released in 1989 has already been accepted by Australia, Canada, South Africa, Western Europe and the USA as a basis for ore reserves reporting standards [12]. In some other East-European countries local standards (mostly originated from the former Soviet's levels of orebody recognition) are being reviewed in order to meet the widely approved "Western" – JORC standards which is necessary for opening the domestic mining industry for foreign investors.

In general, the JORC Standard relies rather on the engineers' skills (the "Competent Person") and the needs of precise documentation of the methods used for reserves identification than on strictly predefined "rules" of ore classification. Following the JORC code, the mineable lignite reserves can be identified when the detailed mine planning has been applied to measured and indicated resources, which allow a reliable estimate of coal seam thickness, quality, depth and tonnage. The term "coal reserves" denotes that part of resources which could be mined economically with regard to realistic conditions (both mining and economic) at the time of reporting (applying known mining technology, costs and pricing data).

Therefore, in order to identify such defined mineable lignite reserves at least the basic plan of an open pit mine has to be built. This is a new approach in comparison to established methods (in Poland and other countries) of identification of mineable lignite reserves.

The methods (like these described in Geology and Mining Law – Act 1994) define the mineable lignite reserves as the resources of lignite seam that keep within preset limits the following parameters:
- minimum thickness of the coal seam;
- maximum stripping ratio;

- maximum depth of the coal seam;
- minimum weighted average calorific value, and
- maximum weighted average total sulphur content of coal.

It must be underlined that all these constraints have been set upon the experience derived from the years of operation of lignite mines with the use of surface mining technology and – in general – such defined area of mineable lignite reserves is not too far from the area of the final plan of an open pit mine. However these constraints have been set on the basis of traditional, manual, plan-and-section, paper based design technology regardless on the available computer technology which uses three-dimensional design environment. The replacement of the former 2-D methods with the new, fully spatial design tools for more detailed identification of mineable lignite reserves is the obvious step.

The methods of computer based, three-dimensional orebody modelling have been successfully developed over the years and have led to two types of models:
- models of structural surfaces that represent the borders of geological strata and topography – open Digital Terrain Models (also known from GIS technology) or closed (solid) models (also known from mechanical design) built by sets of irregular triangles (so called wireframes);
- models representing the spatial distribution of examined ore parameters (quality, geotechnical and others) – usually block models – sets of rectangular prisms which fill the space created by structural models.

The block models which contain the quality data of an orebody give an opportunity to develop (with the use of known mining costs and ore prices) the economic block model of an orebody – the model whose each block has been assigned with its mining cost and – in case of an "ore" block – calculated revenue from selling the contained ore.

Then, after applying mining constraints: the allowed slope angle (depending on the geotechnical parameters and mining requirements – e.g. access road space) and limits-boundaries (if exist) of the final pit the optimisation algorithms that find the ultimate pit can be used. The ultimate pit is the pit that delivers maximum cash flow for given economic block model with regard to mining constraints (mostly slope angle). Though several various algorithms have been developed, the Lerchs-Grossmann graph theoretic algorithm [9] is the only mathematically rigorous method that solves the problem. The obtained ultimate pit identifies this part of resources that are ore reserves following the mentioned above meaning of this term.

These economic optimisation models have been applied mostly for mining metallic mineral deposits where the valuation formula can be obtained on the basis of grade and pure metal price. However while mining lignite, which quality depends rather on blending the raw material to match the requirements of a power plant and there is no market price for lignite at all, the valuation formula is not obvious.

Lignite is extensively used in Poland and several other European countries for power generation. Usually one or more closely located to each other lignite open pits supply one power plant designed for the specific fuel parameters. While several pits are being advanced, some lignite deposits are evaluated. The proper evaluation of their reserves requires the implementation of the newest available optimisation methods.

Instead of grade, the quality parameters have then to be used to build the valuation formula of a non-grade raw material like lignite [10]. If the price formula incorporates all necessary quality parameters that have to be controlled, then the economic optimisation can also lead to obtain the expected blending solution. Some proposals of the lignite price formula are given below.

In the paper [10] economic block model of a lignite deposit has been built with the use of two different formulations to calculate a block net value (BNV):

$$BNV = V \cdot L \cdot UW \cdot P - V \cdot C \qquad (1)$$

$$BNV = \{[(V \cdot L \cdot UW) \cdot (1 - M)] \cdot (1 - A)\} \cdot P_C - V \cdot C \tag{2}$$

where: V – block volume (m³), L – lignite content in block (%), UW – lignite density (assumed 1.2 (t/m³)), P – lignite price at the mine site ($/t), M – moisture of lignite (%), A – (Ash + CO_2) content of dry lignite (%), P_C – "Clean" lignite price ($/t), C – Mining cost ($/m³).

The first formula does not depend on the lignite quality while the second tries to incorporate some quality parameters but requires the non-existing price of the "clean" lignite regardless of the real requirements of the power plant.

In the study [7] the economic block model of the lignite deposit was obtained with the use of the price formula proposed for lignite by the Mining Committee of the Polish Academy of Sciences [4].

In the formula the price of lignite is a function of the three main quality parameters: calorific value, ash content and sulphur content as follows:

$$C = C_B \times \left[1 - \frac{Q_B - Q_R}{6724} - \frac{A_R - A_B}{57} - \frac{S_R - S_B}{10} \right] \tag{3}$$

where: C – actual selling price of lignite, C_B – basic price (price of the standard lignite), Q_B – calorific value of the standard lignite (kJ/kg), Q_R – actual calorific value (kJ/kg), A_B – ash content in the standard lignite (%), A_R – actual ash content (%), S_B – sulphur content in the standard lignite (%), S_R – actual sulphur content (%).

This formula is more quality parameters oriented than the previous ones but the basic values of the standard lignite have been set for the whole lignite mining industry in Poland regardless on the actual requirements of the local power plant, which can vary significantly.

In formulations (2) and (3) (and also in some other formulations used in Poland between mines and power plants) the price formulation of lignite looks as follows:

$$C = C_B \times Q_I \tag{4}$$

where: C – actual selling price of lignite, Q_I – coal quality indicator (based on selected lignite quality parameters), C_B – base price (price of reference or standard lignite).

In order to make the lignite price the proper tool for further blending optimisation, the quality indicator should be set either for each lignite deposit (where the standard lignite means the averaged lignite of the analysed deposit) or for the power plant with regard to the requirements of optimising the efficiency of power generation. Such quality indicator should not be further changed as it represents the physical properties of the lignite and the process of generating power.

The other factor of the formulation – the price of standard lignite represents the average fuel cost of a power plant, so it depends on the energy market rather and should be resulted from the analysis of the profitability of the whole power supply company consisting of a power plant and the supplying lignite mine.

2. COMPARISON OF METHODS OF MINEABLE ORE RESERVES IDENTIFICATION

The still existing method of identification mineable coal reserves, based on checking the limits of selected parameters has the following disadvantages:

- it is tedious as it requires a lot of manual work; even if former paper maps have been replaced with the digital maps and, say, GIS environment;
- it is static – does not include local conditions (e.g. the possibility of blending lignite from several pits in order to meet the power plant requirements – poor lignite still could be useful if blended with the better one and then should not be rejected from the mineable reserves, the classification of coal should follow the power plant requirements which could be changed if it is modernised);
- it depends on criteria used in traditional, two-dimensional design technology based on maps and sections only;
- it does not refer to changeable economic conditions.

The only advantage of this method is its simplicity and a lot of experience among mining engineers who have used to work in such a way for years.

The scheme of the existing method is presented on Figure 2.

The proposed method of identification mineable ore reserves, based on digitally generated ultimate pit with respect to mining and economic conditions provides the following advantages:
- it is quick as the most tedious job – the identification of an ultimate pit is done with the use of computerised algorithm working in the three-dimensional environment;
- it is dynamic – local conditions can be included (e.g. the price of the poor lignite can be recalculated with blending formula to take advantage from the possibility of blending lignite from several pits in order to meet the power plant requirements);
- it depends on fully spatial distribution of an orebody parameters and mining constraints (like slope angles, boundaries of preserved zones within the vicinity of an ultimate pit, etc.);
- it can refer to changeable economic conditions – the ultimate pit design can be easily recreated if economic conditions are changed.

The scheme of the proposed method is presented on Figure 3.

Both schemes have been drafted with the use of the standard for Integration Definition for Function Modelling (IDEF0) which is widely used for projects that require a comprehensive modelling technique for the analysis, development, documentation and/or acquisition of information systems (Fig. 1). This standard, once developed for the US Air Force Laboratories (in 1981) has been then adopted by (but not restricted to) software companies for the purpose of design large information systems with many inputs, outputs and controls that have to interfere. As the whole procedure of mineable coal reserves identification is really complex, this standard is a suitable tool for its presentation.

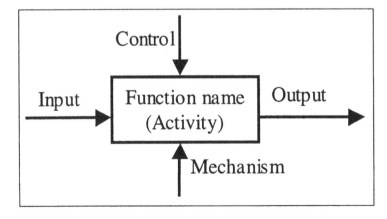

Figure 1. Box and arrows positions and roles in a typical IDEF0 scheme

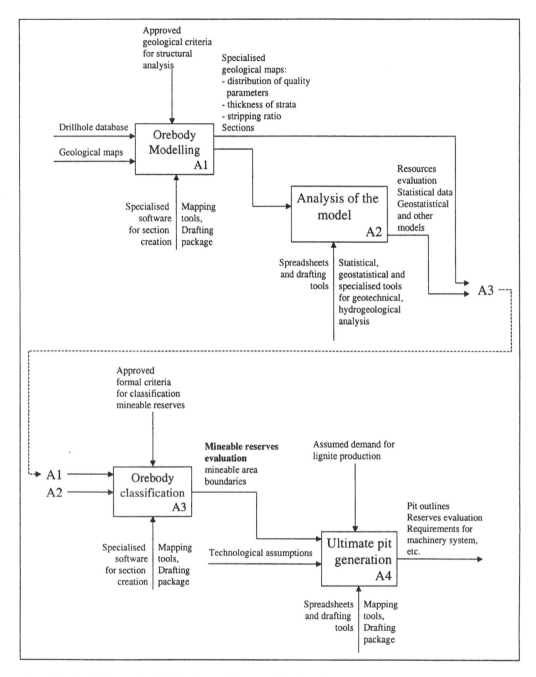

Figure 2. The existing method of mineable coal reserves identification

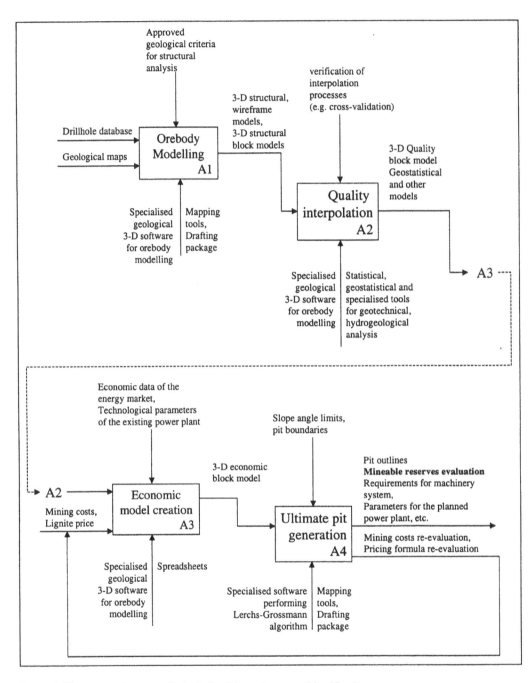

Figure 3. The proposed, new method of mineable coal reserves identification

3. APPLICATION OF THE MODIFIED BILATERAL MONOPOLY SOLUTION
   TO MINEABLE RESERVES IDENTIFICATION

Opencast lignite/coal mine and power plant can be treated as the ideal example of a bilateral monopoly market. Upstream monopoly (a mine) sells intermediate products (lignite) to downstream monopoly (a power plant), which sells final products (electric energy) to end users. In output market a power plant may be a price maker (monopoly) or a price taker (a competitive firm).

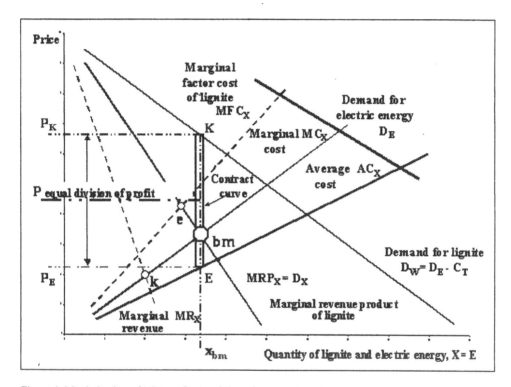

Figure 4. Maximisation of joint profits in a bilateral monopoly [5]

In the classical solution, described in literature (e.g. [3]), price of an intermediate product (lignite) would not have the influence on the choice of produced quantity, which would be determined
based on knowledge of marginal cost of lignite production MC_X and marginal revenue product of
lignite MRP_X. So the lignite price should be negotiated somewhere between prices p_E and p_K (break-
even prices for a mine and a power plant (Fig. 4)) on the contract curve. The negotiated price would
determine the division of profit between both parties showing their bargaining power.
   The classical solution is not the appropriate one in the case of lignite mine and power plant in
long run due to specificity of opencast mining industry.
   In opencast mines due to possibility of application of the Lerchs-Grossman pit optimization technique to mineable reserves estimation there is the direct influence of lignite prices on the amount of
lignite inside the optimal ultimate pit and its phases. This dependency is determined through parameterisation process (a kind of sensitivity analysis of lignite ultimate pit size on change of lignite base
price [8] and has the form of monotonically ascending discontinuous curve (Fig. 5).

This dependency can be treated as the long run lignite supply curve $x = x(p_n)$. It has been shown [5] that because of it the modified bilateral monopoly of lignite mine and power plant, as oppose to the classical case, has the determinate solution not only in the quantity of intermediate product (lignite) but also in its price. We can find the optimal lignite price p_{Vmax} with corresponding optimal ultimate pit which maximises joint profit of the whole bilateral monopoly in long run (its non-discounted value Π_{Vmax}) taking into account lignite deposit and its quality, economic conditions, costs structure, future demand for electric energy and its prices. Amount of lignite x_{Vmax} (Fig. 6) inside this optimal pit can be treated as the mineable reserves adequate for the whole bilateral monopoly of lignite mine and power plant.

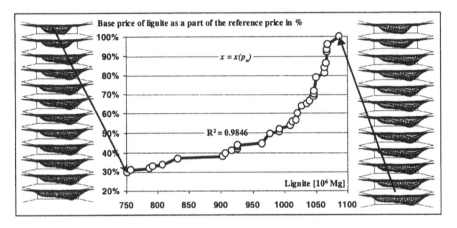

Figure 5. The long run lignite supply from the mine to power plant – the relation between lignite price (presented as a part of the reference price in %) and the optimal ultimate pit maximising non-discounted net cash flows (presented as the amount of lignite inside pit) for 34 nested pits generated for the "Szczercow" deposit [6]

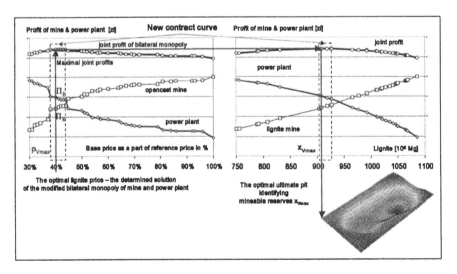

Figure 6. Joint profit of mine and power plant as the function of lignite quantity (for the electric energy price $p_e = 0.13$ PLN/kWh) – the hypothetical solution for the pit placed on the "Szczerców" deposit

The calculated optimal lignite price p_{Vmax} determines the division of maximal profit Π_{Vmax} between the mine Π_K and the power plant Π_E ($\Pi_{Vmax} = \Pi_K(p_{Vmax}) + \Pi_E(p_{Vmax})$) (Fig. 6), what in fact excludes price negotiation, unless:

a. An area around the maximum of joint profit is almost flat or there are a few local maximums on the same level and it is possible to depict the new contract curve (Fig. 6), and

b. During negotiation, in conditions of cooperation and mutual trust, other split of joint profit has been decided and realized through:
  – side payments (in order to decrease the profit levels differences or realize agreed division);
  – determination of other lignite price p_{wt} (the transfer price), which will be used only for clearing accounts between both sides of bilateral monopoly in order to attain agreed profit division ($\Pi_{Vmax} = \Pi_{Kagreed} + \Pi_{Eagreed}$).

In such situation both the selection of ultimate pit as well as other decisions about its shape and size change should be done based on optimal price maximizing joint profits of the whole system in order to keep economic rationality and effectiveness of mine activity. Usage of transfer prices other than optimal requires close cooperation, so it is proposed to treat lignite price negotiation as a two-stage, two-person, cooperative, non-zero, variable sum game [6]. In the first stage (cooperative) both sides should select the ultimate pit maximising joint profits Π_{Vmax} and in the second one (competitive) the agreement should be achieved regarding profit division and calculation of the transfer pri-ce of lignite, which leads to it.

Analysis of this game leads to conclusion that only full vertical integration of mines and power plants can secure realisation of the optimal solution – excavation of the optimal ultimate pit maximising joint bilateral monopoly profits. If the transfer price p_{wt} is lower than the optimal price p_{Vmax} the mine, behaving opportunistically, can adjust its activity to it finding out and excavating the pit, which is optimal only for itself ($\Pi_K(p_{wt}) > \Pi_{Kagreed}$). It means the excavation of a smaller pit than the pit optimal for the whole bilateral monopoly, what is not effective in Pareto sense. It lowers the joint profits, shortens the time of lignite excavation (energy production) and lowers the level of deposit utilization.

CONCLUSIONS

Proper identification of mineable lignite reserves (Fig. 3) should be based on the optimal solution for the modified bilateral monopoly of a power plant and a mine, which utilises pit optimisation techniques.

Due to the inherent conflict of individual and group rationality implicated by the mine information advantage about the deposit and its predominant strategy to optimise only its own profit only the full vertical integration of both sides can secure the Pareto optimal solution.

Such integration does not create any threats to energy market and consumers due to the pit optimal for bilateral monopoly is greater than optimal only for the mine, what should increase lignite and energy supply in long run.

REFERENCES

[1] Announcing the Standard for Integration Definition for Function Modelling (IDEF0). 1993, Draft Federal Information, Processing Standards Publication, No. 183, 1993, December 21.
[2] Act "Geology and Mining Law", 1994 (with later amendments). Diary of Acts of the Republic of Poland.
[3] Blair R.D., Kaserman D.L. & Romano R.E. 1989: A Pedagogical Treatment of Bilateral Monopoly. Southern Economic Journal, No. 55, pp. 831–841.

[4] Blaschke W. 1998: Problems of Lignite Prices with Regard to Pricing Foundations of Coal for Power Engineering. Proceedings of the Session of the Mining Committee of the Polish Academy of Sciences, Krakow–Belchatow (in Polish).

[5] Jurdziak L. 2004: Tandem: Lignite Opencast Mine & Power Plant as a Bilateral Monopoly. Mine Planning and Equipment Selection. A.A. Balkema Publishers, Taylor & Francis Group.

[6] Jurdziak L. 2006: Lignite Price Negotiation Between Opencast Mine and Power Plant as a Two-Stage, Two-Person, Cooperative, Non-Zero Sum Game. International Symposium Continuous Surface Mining ISCSM, Aachen 2006, Department of Mining Engineering III, RWTH Aachen University, DEBRIV Bundesverband Braunkohle.

[7] Jurdziak L., Kawalec W. 2000: Optimisation of the Pit Based on the Price of Lignite and Quality Requirements for the "Szczerców" Deposit. VII Conference Proc., The Exploitation of Mineral Resources, Zakopane 2000 (in Polish).

[8] Jurdziak L., Kawalec W. 2004: Sensitivity Analysis of Lignite Ultimate Pit Size and its Parameters on Change of Lignite Base Price (in Polish). Mining and Geology VII, Scientific Papers of the Institute of Mining Engineering of the Wroclaw University of Technology, No. 106, Studies and Research, No. 30, Wroclaw 2004.

[9] Lerchs H., Grossmann I.F. 1965: Optimum Design and Open Pit Mines. Transactions, C.I.M, Vol. LXVIII, pp. 17–24, 1965.

[10] Mastoris J., Topuz E. 1995: Modelling, Optimization and Sensitivity Analysis of the Final Pit Limits for a Lignite Deposit. Mining Engineering, November, 1995.

[11] Samuelson W.F., Marks S.G. 1998: Managerial Economics. (Polish edition), PWE, Warszawa 1998.

[12] Stephenson P.R. 2003: The JORC Code – Maintaining the Standard. The CIM Mining Conference and Exhibition, Montreal 2003.

[13] Stephenson P.R., Glasson K.R. 1992: The History of Ore Reserve Classification and Reporting in Australia. The AusIMM Annual Conference, Broken Hill 1992.

Publication sponsored by the Foresight Project "Scenarios of the Technological Development for the Lignite Mining and Process Industry" No. WKP_1/1.4.5/2/2006/4/7/585/2006 under Sectorial Operational Program – Increase of Competitiveness of Companies (2004–2006).

International Mining Forum 2008, Sobczyk & Kicki (eds) © 2008 Taylor & Francis Group, London, ISBN 978-0-415-46126-9

Fuzzy Modelling of an Ultimate Pit

Witold Kawalec

Institute of Mining Engineering, Wroclaw University of Technology, Poland

ABSTRACT: An idea of fuzzy modelling of an ultimate pit especially suitable for lignite deposits has been presented. For a lignite deposit various economic models that are dependent on the arbitrary choice of the lignite price formulae can be created. On a contrary to a standard grade based price formulae these formulas can change the classification of the deposit and – consequently – significantly change the Life-Of-Mine Plan. The fuzzy modelling of an ultimate lignite pit built with the use of the standard Lerchs-Grossmann algorithm for the various, alternative price formulas has been implemented as a tool for economic evaluation of the lignite deposit. The case study of the ultimate pit created for the small lignite deposit has been shown.

1. INTRODUCTION

Generation of an ultimate pit with the use of any of the existing, computerised optimisation algorithms (practically Lerchs-Grossmann or – less popular – "floating-cone") have been successfully implemented into the mine planning procedures for many years [7]. However, this method has been widely adopted to metallic ore mining rather while planning of continuous surface lignite mines still relies on traditional analysis of geological maps and sections despite some attempts of implementtation the optimisation methods in this area [8], [4], [9], [10], [6].

There are several reasons for the visible delay of acquiring of new computerised optimisation methods by continuous surface lignite mines. In many countries energy prices have been controlled by the government or even the lignite mines together with the mine-mouth power plants have been nationally owned which have not created the pressing on improving their economical efficiency. Lignite mines incorporating continuous surface mining system favour long mining panels which leads to the their lack of flexibility (in comparison to cyclical mining systems). The third and probably the most important obstacle is the fact that lignite is not offered on a free market and its in-situ price (that is necessary for creating the economic block model of lignite deposit) cannot be obtained upon the analysis of stock exchange data like metal prices. Within RWE Power AG – one of the top rated European lignite mining & power generation company – no lignite price has been set even for internal accounting though the large set of lignite quality parameters is precisely measured and controlled in order to meet strict power plants or other processing plants requirements [4].

In Poland several formulas of lignite price have been developed throughout the years either during negotiations between lignite mines and power plants or as a result of academic research [2], [Grudziński 1999]. All solutions have been based on the idea of the use of lignite quality parameters to build the valuation formula of a non-grade raw material [8]. Though such approach seems to support the need of quality control of the fuel that is converted into energy no hitherto lignite price formula seems to meet all power plant requirements.

The general price formulation of lignite looks as follows:

$$C = C_B \times Q_I, \tag{1}$$

where: C – actual selling price of lignite, Q_I – coal quality indicator (represents the "grade" of coal, a function of quality parameters), C_B – base price (price of reference or standard lignite).

Both the base price and the coal quality indicator can be changed following the negotiations between a lignite mine and a power plant (if these companies are separate), following internal accounting needs (within a single company) or following the changes of interpretation of coal quality parameters for the purpose of producing energy. It must be pointed that – on a contrary to standard grade in case of metal ore deposits – any replacement of a coal quality indicator with another one can alter the relative value of "ore" blocks in the economic block model of the deposit. Block A can be worth less or more than block B depending on the assumed function of the coal quality indicator. Therefore alternate price formulas change the spatial distribution of the coal value that is modelled in the economic block model which eventually leads to different ultimate pit models defining mineable lignite reserves of a analysed deposit. Consequently, these different models do not create a set of nested pits (a product of changes of a base price) but rather a set of hybrid pits [11] which require further processing to build a final solution – a single ultimate pit. The method of processing of the obtained set of ultimate pits generated for various lignite price formulas has been described below.

2. FUZZY MODELLING APPROACH

A solution can be found with the use of fuzzy sets. A set is called a fuzzy set if its characteristic function (also known as membership function) can get any value from the interval [0,1] – not only 0 or 1 like in the case of non-fuzzy (crisp) sets. Since the fuzzy set theory was presented, it has been widely used when neither the exact values nor the random distribution of given parameters are known. Fuzzy description is very common – people say "some 5 tonnes" which can mean either 4,80 tonne or 5,10 tonne or exactly 5,0 tonne.

Unexpectedly, despite the fact that any geological interpretation and orebody evaluation is imprecise, fuzzy logic has been implemented there just recently [3]. More popular among geologists have been random models like conditional simulation for creating the equally probable orebody block models upon the raw drillhole data and geostatistical analysis. These equally probable orebody models are the input data for generating the equally probable ultimate pits – already mentioned hybrid pits. The standard Boolean operations of hybrid pits like union or intersection have already been analysed [11] and it has been proved that the output models (an union or a intersection of a set of hybrid ultimate pits) maintain all required parameters of the input models (mainly slope angles) and therefore can be treated as the final ultimate pits. This result is very important for the possibilities of wider implementation of set operations for generating an ultimate pit. More general solution of the problem described in the introduction has been solved with the use of fuzzy modelling procedure adopted to the standard Lerchs-Grossmann ultimate pit generation.

2.1. *Fuzzification of an ultimate pit for a given price formula*

Lerchs-Grossmann (L-G) algorithm processes a three-dimensional economic block model of the deposit – the model whose each block has been assigned with its mining cost and – in case of an "ore" block – calculated revenue from selling the contained ore.

Let X be a finite set of all cells of orebody and overburden block model.

For the purpose of the L-G algorithm all blocks (or cells) share identical dimensions (no subcells allowed) and can be easily indexed as a three-dimensional matrix:

$$X = \{x_{ijk} : i \in I, j \in J, k \in K\} \tag{2}$$

where: I, J, K – finite set of indexes of cells x.

Each cell x – an element of the set X – holds individual values of considered parameters of the deposit (quality, geotechnical, etc.) $p_1, p_2, ..., p_m$.

An ultimate pit of the deposit is a finite subset of the space X, generated by L-G algorithm with regard to price and costs formulas and constraints rules:

$$UP = LG(X, C, K, L) \tag{3}$$

where: $C = c(x_{ijk})$ – price formula, $K = k(x_{ijk})$ – cost formula, $L = l(x_{ijk})$ – pit constraints (slope angle, boundaries, etc.).

Assume that the layout of a price formula follows (1) and the quality indicator is a function of parameters $p_1, p_2, ..., p_m$. The base price C_B and quality indicator Q_I are independent form each other.

Hence:

$$UP = LG(X, C_B, Q_I, K, L) \tag{4}$$

Usually, apart of generating a single ultimate pit for a given base price a set of nested pits (phases): $UP_1, UP_2, ..., UP_t$ (where UP_t is the smallest pit) generated for a series of discounted base price: $(1-\beta)C_B, (1-2\beta)C_B, ...$, where: β is a base price discount level $(1 \div 5\%)$.

Following (4):

$$UP_1 = LG(X, (1-\beta) \cdot C_B, Q_I, K, L); \quad UP_2 = LG(X, (1-2\beta \cdot C_B, Q_I, K, L) \tag{5}$$

The nested pits $UP_1, UP_2, ..., UP_t$ are not equally probable (like hybrid pits) because their discounted base price value is less probable than the assumed base price. Therefore it should also be treated as a fuzzy number – an example of a fuzzy set [12]. The shape of its membership function (see a sample one on figure 1) should be created upon the detailed analysis of market trends (e.g. trends of energy price which influence the lignite base price). Such fuzzyfied base price reflects a prediction of an actual base price.

The bottom figure shows the sample chart of a membership function of a fuzzy number base price, full memberships (function value equals 1) has been set to one point only: C_B – the expected (or the preset) value of the base price. The lower (as well as the greater) values of the base price have obtained smaller values of the membership function (according to assumptions derived from the external analysis). However, the shape of the membership function on Figure 1 is purely subjective and in each real case it has to be a result of serious investigations.

The decreased level of membership of the fuzzy number base price that are adjusted to values of the base price that are smaller than the preset value C_B can be used for weighting the nested ultimate pits (phases) generated for these values.

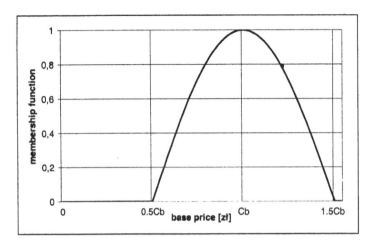

Figure 1. Sample chart of a membership function of a fuzzy number base price

These weighted nested pits eventually build a fuzzy ultimate pit with a membership function that is computed from the equation:

$$UP_{fuz}(x_{ijk}) = \frac{UP(x_{ijk}) + UP_1(x_{ijk}) \cdot \mu_{c_B}((1-\beta)c_B) + .. + UP_t(x_{ijk}) \cdot \mu_{c_B}((1-t\beta\beta)_B)}{1 + \mu_{c_B}((1-\beta)c_B) + .. + \mu_{c_B}((1-t\beta\beta)_B)} \qquad (6)$$

where: $UP(x_{ijk})$ – membership function of an non-fuzzy (crisp) set UP (equals 0 or 1), μ_{cB} – membership function of a fuzzy number base price (see Fig. 1).

It has to be checked whether the equation (6) meets the requirements of the fuzzy set theory to be considered as a proper definition of a fuzzy set membership function.
It is easy to find that:
a) All values of the function $UP_{fuz}(x_{ijk})$ are non-negative.
b) The maximum value of this function equals 1 (only for x_{ijk} that belongs to the ultimate pit and all phases UP, UP_1, UP_2, ..., UP_t).
c) If x_{ijk} does not belong to any of the phases UP, UP_1, UP_2, ..., UP_t than $UP_{fuz}(x_{ijk}) = 0$ equals 0.
d) If x_{ijk} belongs to any of the phases than $UP_{fuz}(x_{ijk}) \in [0,1]$.
The set UP_{fuz} is the fuzzy set, created with the fuzzification operation made on non-fuzzy (crisp) sets UP, UP_1, UP_2, ..., UP_t with the use of formula (6) with regard to the preset membership function base price.
Elements that belong to all phases create its core:

$$core(UP_{fuz}) = \{x_{ijk} \in X : UP_{fuz}(x_{ijk}) = 1) \qquad (7)$$

Note that $core(UP_{fuz}) = UP_t$.
The support of the UP_{fuz} fuzzy set is defined as:

$$supp(UP_{fuz}) = \{x_{ijk} \in X : UP_{fuz}(x_{ijk}) > 0) \qquad (8)$$

Again, note that supp(UP_{fuz}) = UP – the ultimate pit generated for base price = C_B.

The fuzzy set ultimate pit UP_{fuz} strongly depends on the construction of the membership function of a fuzzy number base price.

2.2. Aggregation of fuzzy ultimate pits generated for various price formulas

The fuzzy ultimate pits generated for various ultimate pit formulas have to be aggregated in order to build their final product – the all-formulas (AF) fuzzy ultimate pit. Within the fuzzy set theory various operators are for creating aggregation of fuzzy sets: intersections and unions. The MAX (for union) and MIN (for intersection) operators are the most common [12].

The membership function of the intersection of fuzzy ultimate pits (each generated for its price formula f_s) can be set as the minimum value of membership levels of all pits considered:

$$\bigcap_s UP_{fuz}(x_{ijk}) = \min_s UP_{fuz}(f_s)(x_{ijk}) \tag{9}$$

while the membership function of the union of fuzzy ultimate pits can be set as the maximum value of membership levels of all pits considered:

$$\bigcup_s UP_{fuz}(x_{ijk}) = \max_s UP_{fuz}(f_s)(x_{ijk}) \tag{10}$$

The intersection and the union of a series of fuzzy ultimate pits provide the bottom and the top constraints of its real borders. It is recommended to generate both pits for further processing.

In the case of diversified level of importance of various price formulas the weighted operators should be used instead of (9) and (10) [12].

2.3. Defuzzification of aggregated fuzzy ultimate pits

The aggregated fuzzy ultimate pits can be used for various analysis (like the level of their fuzziness) but for any further planning activities the crisp – not fuzzy – ultimate pit is necessary.

The crisp set can be obtained with the use of typical defuzzification operator: α cut crisp set:

$$UP_\alpha = \{x_{ijk} \in X : A\,UP_{fuz}(x_{ijk}) > \alpha\} \tag{11}$$

where: $AUP_{fuz}(x_{ijk})$ – membership function of an aggregated (intersection or union) fuzzy ultimate pit.

The value of "cutting" the fuzzy set – α should be chosen upon the examination of the aggregated fuzzy ultimate pits. As already mentioned the $UP\alpha$ ultimate pit meets all geometrical constraints (pit limits, slope angles) that have been preset to the ultimate pit generation.

The whole procedure of fuzzification of a set of the ultimate pits with phases, aggregation the fuzzy ultimate pits and their final defuzzifiaction not only brings the single, mathematically supported solution that otherwise would be questionable. Its biggest advantage is the large amount of information about the sensitivity of the ultimate pit model on the variations of both price formulas and the base price. The analysis of fuzziness made for chosen sectors of the ultimate pit can be used for better (in terms of economic optimization) decisions of detailed pit design: haul roads location, pit advancing direction or even changing of the ultimate pit shell with regard to changes of economical environment.

2.4. *Next steps*

No model should be treated as the optimal one and the best ever. The proposed above procedure of generating the ultimate pit with the use fuzzy modelling applied to various price formulas and base price can be extended to different cost formulas. Consequently, the final ultimate pit can be produced from a set of many fuzzy sets each representing an ultimate pit generated for given price and cost formulas.

The idea of generating hybrid pits for the set of economic block models obtained from conditionally simulated quality block models has already been implemented into the premier open pit optimization software – NPV Scheduler (released by Datamine Corp.). This shows that there is a growing tendency towards more sophisticated processing the geological and mining data in order to get a better understanding of a mining project.

3. CASE STUDY

The generation of the ultimate pit for a small lignite deposit "Morzyczyn" of the opencast lignite mine "Konin" in central Poland has been done following the presented method. Due to geological documenttation the deposit contains some 9,2 mln tonnes of mineable reserves (category C-1 according to Polish geology & mining law), average thickness of lignite seam is 5,7 m and of overburden is 48 m. The lignite seam is split with inter-layers.

The coal quality parameters are below the power plant requirements:
a) Calorific value: avg. $Q_R^i = 7000$ kJ/kg (required minimum: 7800 or 8300 kJ/kg).
b) Sulphur content: avg. $S_R = 0,94\%$ (allowed maximum: 0,96%).
c) Ash content: avg. $A_R = 18\%$ (allowed maximum: 12%).

High stripping ratio as well as the close neighbourhood of the protected landscape area are the next arguments against the exploitation of this deposit. However, even so poor deposit could be considered as subsidiary source of lignite which can be blended with the better one exploited in another pits of the multi-pit "Konin" mine to provide either the bigger annual capacity of the whole mine or the extension of its lifetime. The sample analysis of the Life-of-Mine plan built upon the ultimate pit of the "Morzyczyn" deposit generated for the price formula calculated for the blended coal has been presented in [5]. The analysis showed that the multi-pit approach has increased the value of the poor deposit and – consequently – enlarged the mine reserves.

3.1. *Quality modelling of the deposit*

The orebody modelling has been done in the Datamine Studio geological & mining software. The structural wireframe models of the terrain, top and bottom lignite seam surfaces and waste interlayer zones have been interpreted from the raw drillhole data. Then the structural block model has been built with the cells sizes of 120×120×4 m (see Fig. 2). The model contains three main zones: coal, overburden and interlayers.

Due to relatively low density of drillhole grid (350×350 m over the area of some 30 km^2) interpolation of quality parameters has been done with the use the simplest "nearest neighbour" method. The total tonnage of the rocks within the "coal" zone has reached 85 million tonnes (regardless of actual coal seam thickness, stripping ratio and quality parameters). The average values of the calorific value, ash and sulphur content were very similar to those from geological documentation. Figure 2 shows the general plan view of the deposit together with the sample section of the structural block model.

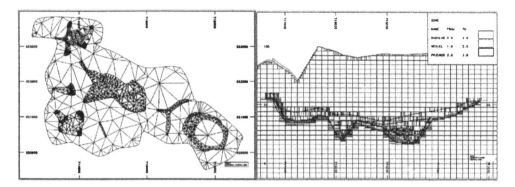

Figure 2. Structural wireframe (plan view – left) and block model with wireframe slices of the lignite deposit with waste interlayers (section with vertical exaggeration – right) [5]

3.2. Price and costs formulas

The quality block model has been used for building the economic block model with two different coal price formulas:

F_1 – the formula used in the "Konin":

$$C = C_B \cdot \left[\frac{Q_R}{8850} - \frac{A_R - 12}{200} - \frac{S_R - 0,6}{10} \right] \tag{12}$$

F_2 – the formula proposed for lignite by the Mining Committee of the Polish Academy of Sciences [2]:

$$C = C_B \times \left[1 - \frac{Q_B - Q_R}{6724} - \frac{A_R - A_B}{57} - \frac{S_R - S_B}{10} \right] \tag{13}$$

where: C – actual selling price of lignite, C_B – base price (price of the reference lignite), Q_B – calorific value of the reference lignite (kJ/kg), Q_R – actual calorific value (kJ/kg), A_B – ash content in the reference lignite (%), A_R – actual ash content (%), S_B – sulphur content in the reference lignite (%), S_R – actual sulphur content (%), $Q_B = 9002$ kJ/kg, $A_B = 7,5\%$, $S_B = 0,6\%$, $C_B = 57,65$ zł/t.

According to the idea of blended coal from the multi-pit mine and setting to u = 16% the share of "Morzyczyn" output in the whole mine output, the actual values of quality parameters have been replaced with the parameters of blended coal:

$$Q_R = u \cdot Q_R(M) + (1 - u) \cdot Q_{avg}, \ S_R = u \cdot S_R(M) + \tag{14}$$
$$+ (1 - u) \cdot S_{avg}, \ A_R = u \cdot A_R(M) + (1 - u) \cdot A_{avg}$$

where: $Q_R(M)$, $S_R(M)$, $A_R(M)$ – actual parameters from the block model of the "Morzyczyn" deposit, Q_{avg}, S_{avg}, A_{avg} – average values of the quality parameters of coal from three other pits of the "Konin" multi-pit mine that could be mined together with "Morzyczyn".

The simple mining cost value has been assumed upon the analysis of mining costs in the "Konin" mine (5,25 zł/cum) regardless on the depth of mining and other possible adjustment. The average slope angle has been set to 12°.

3.3. Fuzzified ultimate pits for alternative price formulas

On the basis of settings described above the optional ultimate pits have been generated with the use of NPV Scheduler mining optimisation software. The quality block model has been imported into the program, then optional economic models have been created and, eventually, the two variants of the ultimate pit each with a set of nested pits (for the base price discount level $\beta = 5\%$).

The analysis of the obtained ultimate pits (see Tab. 1 and top two pit shells on Fig. 6) shows the strong influence of the price formula (not the base price) onto the final results, especially the mineable lignite reserves in both pits. The first phase of each ultimate pit has been generated within the specific, small part of the lignite deposit under significantly thinner layer of overburden but remotely located from the main coal field and therefore it should be neglected.

Table 1. Parameters of the generated ultimate pit (with phases)

Base price [% C_B]	Ultimate pit for price formula F1		Coal reserves [t] $\times 10^6$	Stripping ratio	Ultimate pit for price formula F2		Coal reserves [t] $\times 10^6$	Stripping ratio
	Phaseprofit [zl]$\times 10^6$				Phaseprofit [zl]$\times 10^6$			
60	1	1,4	0,06	4,6				
65	2	109	5,5	5,4				
70	3	171	9,0	5,5	1	0,9	0,06	4,6
75	4	251	14,0	5,7	2	63	5,4	5,3
80	5	377	25,0	6,3	3	99	9,0	5,5
85	6	438	32,5	6,5	4	133	13,5	5,7
90	7	562	50,0	7,0	5	136	14,1	5,7
95	8	606	63.5	7,3	6	171	25,5	6,3
100	9	610	67,5	7,4	7	174	31,0	6,5

Following the described procedure of fuzzification, the membership function of a fuzzy number base price has been assumed (see Fig. 3).

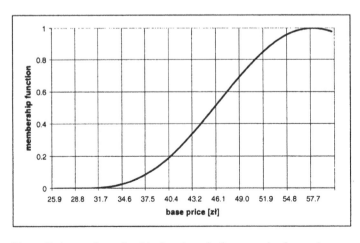

Figure 3. Assumed membership function of a fuzzy number base price

102

The fuzzified two ultimate pits: UP_{F1fuz} and UP_{F2fuz} have been produced in Datamine with the use of advanced spreadsheet and block model processing commands. After fuzzification (6), the apparently very different minaeble reserves of both ultimate pits have been brought closer together with regard to the level of membership of their elements (see Tab. 2).

Table 2. Evaluation results of fuzzified ultimate pits UP_{F1fuz} and UP_{F2fuz}

Membership level of blocks	Fuzzified UP, formula F1		Fuzzified UP, formula F2	
	Coal reserves [t] $]\times10^6$	Overburden volume [cum]$\times10^6$	Coal reserves [t] $]\times10^6$	Overburden volume [cum]$\times10^6$
below 0,3	53,2	405	18,1	124
0,3÷0,7	9,1	50,7	8,5	47
above 0,7	5,5	28,3	5,4	27,7

3.4. Analysis of output models

The fuzzified ultimate pits UP_{F1fuz} and UP_{F2fuz} have been aggregated and the membership function of their fuzzy intersection (9) and fuzzy union (9) have been calculated for each cell in the output model. After investigation of the histogram (see Fig. 4) calculated for the support (8) of the fuzzy union the value of cutting the fuzzy set (11) has been set to 0,3. The noticeable near zero frequency of full membership elements is caused by very small dimensions of the first phase. This means that the core (7) of the fuzzy set is very small but in fact the area containing the elements of the near full membership (around 0,8) could be treated as the core. The evaluation results are given in Table 3.

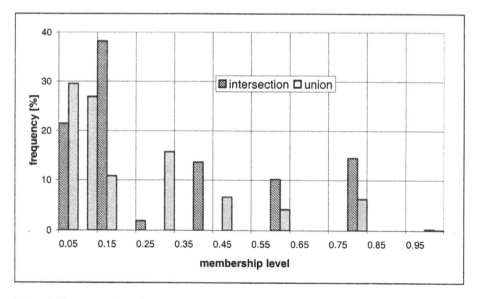

Figure 4 Histogram of membership level of the fuzzy intersection and fuzzy union of aggregated fuzzified ultimate pits

103

Table 3. Evaluation results of output models

| Membership level of blocks | Fuzzy, α cut intersection | | Fuzzy, α cut union | |
	Coal reserves [t]]×10⁶	Overburden volume [cum]×10⁶	Coal reserves [t]]×10⁶	Overburden volume [cum]×10⁶
below 0,3	18,0	125	55,2	405
0,3÷0,7	8,5	47,2	9,1	50,7
above 0,7	5,4	27,7	5,5	28,2
α cut crisp set	13,8	74,6	14,4	78,3

The final α cut crisp set – the new ultimate pit, developed in Datamine has been read back into NPV Scheduler for detailed economic investigation. One of the capabilities of the procedure of generating the ultimate pit in the NPV Scheduler is the preliminary assessment of the NPV of the exploitation of an ultimate pit, made on the basis of the presumed annual output and annual discounting. The "lookahead" NPV of each block included into the ultimate pit is computed with regard to slope angle. Then all blocks are sorted to provide the NPV Optimal Extraction Sequence (OES) – the sequence that delivers the highest NPV for a given economic model and general slope angles. Though OES is generated for the purpose of generating the Life-Of-Mine schedule rather and should not be treated as a real plan of mining, the comparison of the highest NPV obtained for input ultimate pits and their final output α cut crisp union (Fig. 5) shows how the proposed solution "averages" the different, input models. Figure 6 presents the visualisation of these ultimate pit shells which identify the area of the most valuable coal reserves regardless on the used price formula.

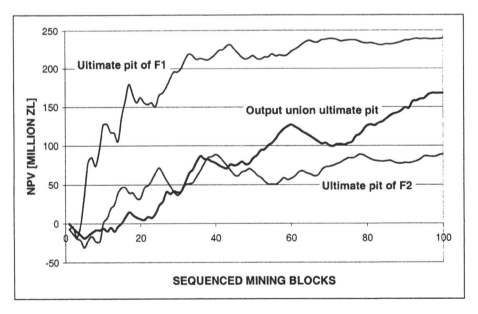

Figure 5. The top limit of the NPV for the input F1 and F2 ultimate pits
and the output α cut union UP computed for Extraction Sequence Optimized for NPV

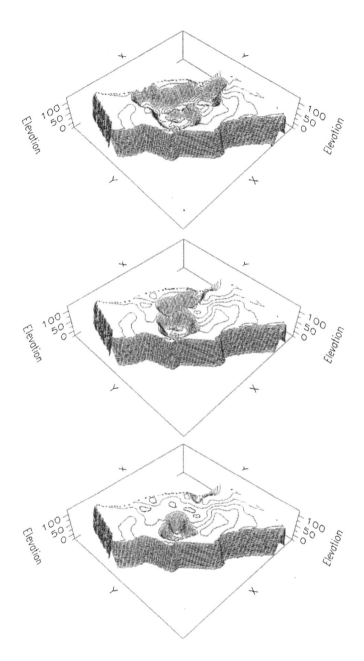

Figure 6. Visualisation of ultimate pit shells. From top: the input F1 UP, F2 UP and the output α cut union UP

CONCLUSIONS AND REMARKS

The presented method of generation the ultimate pit with the use of fuzzy modelling of several Lerchs-Grossmann ultimate pits extends the application of the standard procedure onto the deposits that can be valuated in many ways. Lignite is probably the most important raw material which is not sold on a market and cannot be easily set with an in-situ price like any metal ore. In Poland various price formulas for lignite have already been used which should be taken into account while building an economic model of a lignite deposit. In such a case a set of economic models should be processed to create the final ultimate pit together with the whole analysis of its sensitivity on the input parameters.

The fuzzy modelling approach seems to supplement the newly developed idea of hybrid pits generated upon the set of quality block models obtained with the use of conditional simulation of the input discrete geological data. The more competitive mining industry requires more advanced methods of economic analysis of mining projects and the presented fuzzy modelling of the ultimate pit is an attempt of meeting these growing requirements.

The original data for the presented case study have been used upon the permission of the "Konin" lignite mine. The obtained results are illustrative only and should not be treated as the certified evaluation results.

The computations have been made with the use of the Datamine Studio and NPV Scheduler mining software licensed for the Wroclaw University of Technology by Datamine Corporation.

REFERENCES

[1] Act "Geology and Mining Law" 1994 (with later amendments). Diary of Acts of the Republic of Poland.
[2] Blaschke W. 1998: Problems of Lignite Prices with Regard to Pricing Foundations of Coal for Power Engineering. Proceedings of the Session of the Mining Committee of the Polish Academy of Sciences, Kraków–Belchatow (in Polish).
[3] Demicco R., Klir G. 2004: Fuzzy Logic in Geology. Elsevier Acad. Press, San Diego 2004.
[4] Jurdziak L., Kawalec W. 2000: Optimisation of the Pit Based on the Price of Lignite and Quality Requirements for the "Szczerców" Deposit. Proceedings of the VII Conference: "The Exploitation of Mineral Resources", Zakopane 2000 (in Polish).
[5] Kawalec W., Słowinski J. 2005: Long Term Planning of the Lignite Pit in an Multi-Pit Mine with the Use of Optimisation Methods. Scientific Papers of the Institute of Mining Engineering of the Wroclaw University of Technology, No. 112, Conferences 44, Wrocław 2005.
[6] Kawalec W. & Specylak J. 2000: Open Pit Design Optimisation of a Lignite Deposit. Mine Planning and Equipment Selection, A.A. Balkema 2000.
[7] Lerchs H., Grossmann I.F. 1965: Optimum Design and Open Pit Mines. Transactions, C.I.M., Vol. LXVIII, pp. 17–24, 1965.
[8] Mastoris J., Topuz E. 1995: Modelling, Optimization and Sensitivity Analysis of the Final Pit Limits for a Lignite Deposit. Mining Engineering, November 1995.
[9] Nakara S. et al. 2000: Optimising Extract of Mineral Body by Using Datamine Software. „A Case Study Vastan Lignite Mine – GIPCL". Asian DATAMINE User's Meeting, Udaipur 2000 (not published).
[10] Specylak J., Kawalec W. 1998: Modelling of the Final Pit of a Lignite Deposit with the Use of Lerchs – Grossmann Optimisation Algorithm. Proceedings of the VI Conference: "The Exploitation of Mineral Resources", Zakopane 1998 (in Polish).
[11] Whittle D., Bozorgebrahimi A. 2004: Hybrid Pits – Linking Conditional Simulation and Lerchs-Grossmann Through Set Theory. Symposium: "Orebody Modelling and Strategic Mine Planning", Perth 2004.
[12] Yager R., Filev D. 2004: Essentials of Fuzzy Modelling and Control. John Wiley & Sons (authorised translation published by WNT, Warszawa 1995).

Publication sponsored by the Foresight Project "Scenarios of the Technological Development for the Lignite Mining and Process Industry" No. WKP_1/1.4.5/2/2006/4/7/585/2006 under Sectorial Operational Program – Increase of Competitiveness of Companies (2004–2006).

International Mining Forum 2008, Sobczyk & Kicki (eds) © 2008 Taylor & Francis Group, London, ISBN 978-0-415-46126-9

Energy Policy of Poland in the Beginning of 20th Century in the Opinion Polish Society – Results of the Survey

Alicja Byrska-Rąpała

AGH – University of Science and Technology, Faculty of Management, Cracow, Poland

ABSTRACT: When observing the energy balance structure of the world, European Union or developed countries, such as the USA, Germany or France, one can see the directions of modern power industry. Ecological sources of energy: natural gas, crude oil have become dominant, also, the role of hydroelectricity is increasing. The share of coal – the fuel of the 20th century – is decreasing. Poland is one of the three world's countries, which base their energy industry on coal. The consumption of other primary sources of energy – crude oil, natural gas and hydroelectricity is low, and there are still no nuclear plants built, although works on them were quite advanced in the '80s and the beginning of the '90s. The old-fashioned energy balance structure in Poland is due to low energy consciousness of the Polish society, which is unaware of the significant discrepancy between Polish energy culture and energy policy of the European Union and other countries of the continent.

The paper presents the results of the opinion survey on energy consciousness of the Polish people. The survey concerned consumption structure of primary energy sources in their own country, the European Union and six selected countries. The results of the sample of Polish students were compared with the results of Finnish and Italian students, who had been given an identical questionnaire.

KEYWORDS: Energy consciousness, environment consciousness, sociology, energy sector, coal, crude oil, natural gas, nuclear energy, renewable energy

1. INTRODUCTION

The biggest problem of Polish energy sector is an old-fashioned structure of primary energy sources consumption. Energy sector of Poland is based in an exceptional way on hard and brown coal (such coal dependence occurs only in South Africa and China). In effect, structure of energy consumption in Poland differs radically from that of all developed countries (Table 1). The consumption of other primary sources of energy – crude oil, natural gas and hydroelectricity is low, and there are still no nuclear plants built, although works on them were quite advanced in the '80s and the beginning of the '90s. Question arises – how this situation could develop? Why elected in democratic way representatives of the society (members of parliament and government) did not see the changes occurred abroad and did not modernize the energy sector? Why they during the decades did comply under pressure of the coal lobby and realize the policy of protection of coal lobby's interests with harm for natural environment, state finances and prestige of the country?

When observing the energy balance structure of the world, European Union or developed countries, such as the USA, Germany or France, one can see the directions of modern power industry. Ecological sources of energy: natural gas, crude oil have become dominant, also, the role of hydroelectricity is increasing. The share of coal – the fuel of the 20th century – is decreasing.

Table 1. Structure of primary energy consumption
in selected areas and countries in 2005, percent

Country, region	Crude oil	Natural gas	Coal	Nuclear energy	Hydro-electricity
World	36.4	23.5	27.8	6.0	6.3
OECD	41.0	23.0	21.1	9.6	5.3
USA	40.4	24.4	24.6	8.0	2.6
China	21.1	2.7	69.6	0.8	5.8
European Union	40.8	24.7	17.5	12.9	4.1
Poland	23.9	13.3	61.8	–	1.0
France	35.5	15.4	5.1	39.1	4.9
Germany	37.5	23.9	25.3	11.4	1.9
Hungary	28.1	48.6	10.8	12.5	0.0
Norway	21.7	8.8	1.1	–	68.4
Russia	19.1	53.7	16.4	5.0	5.8
Ukraine	9.9	46.9	26.8	14.4	2.0

Source: BP Statistical Review of Word Energy 2005 (www.bp.com)

Also after change of political system in 1989 no energy sector modernization took place – no nuclear plant was built, gas consumption did not increase, and only change observed was a slight increase in crude oil consumption due to strong development of motorization. Such a situation is due to the long-term pro-coal policy of all the Polish governments, which have complied under pressure of the coal lobby for years. The society has not been informed about the energy revolution in the world by means of both official and unofficial censorship. The old-fashioned energy balance structure in Poland is due to low energy consciousness of the Polish society, which is unaware of the significant discrepancy between Polish energy culture and energy policy of the European Union and other countries of the continent.

Education activities have a kind of feedback – telling the society the truth about energy will not only make strategic decisions easier for the government but also will make the society demand good decisions and, in the end, choose these politicians, who can suggest modern energy solutions.

1.1. *Energy sociology – review of the literature*

The term "energy consciousness", in the literature, appears often and in connection with very different issues. The most popular use of the term relates to projects aimed at energy conservation and improvement of energy efficiency. "Energy consciousness" is connected with heating and air conditioning of buildings, and especially with energy saving investments [10], [13], [14].

The newest trend in energy consciousness development is a strategy of manufacturing products with low energy consumption. The strategy, called Environmentally Conscious Product Strategy, relates to such the products like cars, computers, copying and washing machines, light bulbs etc [11].

The majority of publications dealing with energy consciousness, however, reckon it as part of the broader and more popular concept of ecological consciousness, called "green consciousness". The concept appears in the papers that consider the energy policy from the ecological point of view. The necessity of consciousness integration on all levels of decision making, the long-term effects of decisions made, and the need to increase the role of market forces and competition is underlined [3], [4], [6], [9], [12].

Intensive studies of public opinion on energy issues are carried also by the European Commission [1], [5]. So called Eurobarometers comprise all Member States of the European Union and deal with broad spectrum of energy issues like: general perceptions of energy, the structure of and trends in energy use, priorities in the energy sector, individual behaviour and energy polices [5];

priorities to reduce energy consumption and dependency on imported energy sources, energy consumption habits and willingness to change them [1].

In Poland, the researches of public opinion on energy issues were carried out only when coal industry was restructured and protests of miners had place [2], [7]. All such studies of public opinion shown also that majority of Polish population is against the restructuring of coal industry, treating the process as disadvantageous for the whole national economy and for all Polish citizens. The reform of coal industry was ranked by the respondents on the last place amongst all social reforms done in Poland in last years (health care, pension system, education etc.).

2. RESULTS OF THE STUDY

In the present study, the term "energy consciousness" is used in narrower meaning. It relates to general perception, by the Polish society, of primary energy sources used in Poland and in other European countries. So the aim of the study is to get a people's image of Poland's energy policy as compared with other countries and especially with the European Union.

In order to study the energy consciousness of the Polish society in the terms of perception of primary energy sources used by the European countries listed in Table 1. A public opinion survey was conducted at the beginning of 2006. A representative sample of 880 students from five Polish universities has been taken as well as two small samples: 106 students from North Karelia Polytechnic of Joensu in Finland, 141 students from University of Bergamo in Italy.

Table 2 presents the part of the questionnaire used for the survey, contained questions regarding energy sources.

Table 2. Part of questionnaire used for survey of energy consciousness

What percentage of primary energy consumed comes from...	In your country (Poland, Finland, Italy)	In European Union
• Coal		
• Crude oil		
• Natural gas		
• Nuclear power plants		
• Water power station and other renewable sources		
Total	100%	100%

Respondents were asked numbers as opposite to the standard practice (e.g. in the barometer from 2006, when asking share of individual energy sources, a choice was given between "small", "medium", "large", and "I do not know"). As survey results are estimations, the statistical procedures were used to establish confidence limits for real mean [8]. Considering the limitation of this article, author describes briefly part of results of the survey.

2.1. Results of the Polish people perception study

Information about the perception of different energy sources by Polish students is shown in Table 3. In the table – for each question of the questionnaire – the median value of quantity perceptions and arithmetic mean was given. Moreover, fractions of answers regarded as "correct", "underrated" and "overrated" were shown.

109

Table 3. Statistics describing perception of energy sources shares

Source of energy	Real share (%)	Median	Mean	Fractions of answers		
				Underrated	Correct	Overrated
Coal	Poland (61.1)	60.0	58.39	44	22	34
	EU (18.2)	30.0	29.05	15	27	58
Crude oil	Poland (21.6)	10.0	14.09	64	26	10
	EU (37.5)	20.0	19.64	91	7	2
Natural gas	Poland (11.9)	15.0	17.58	15	36	49
	EU (23.7)	20.0	18.74	47	34	19
Nuclear power plants	Poland (0.0)	0.0	2.67	–	58	42
	EU (14.6)	15.0	17.33	22	38	40
Renewable energy	Poland (5.4)	5.0	7.28	3	83	14
	EU (6.0)	15.0	15.91	1	47	52

Notice: The question – which answers are correct and what part of the respondents has the perception reflecting the reality – was solved by assuming that the evaluation of each source's share is correct if it fits the range of ±7,5% around the real share larger than 25% or the range of ±5,0% around the real share smaller than 25%. This rule does not refer to the share of nuclear energy in Poland and Italy – this evaluation can be either correct (no share) or overrated.

When analyzing the survey results (Table 3) it was stated that one third of Polish students still favour the legend of the mining industry and overrated the share of coal, half of them think that in Poland changes following the world trend of limiting coal consumption have taken place and underrates its share. A completely different phenomenon is observed when analyzing the perception of coal in European power industry. Coal share in the EU is highly overrated, it can be seen that in Polish minds a phenomenon of assimilation (imitation) of two energy cultures – the Polish one, based on coal, and the European one, treating coal as the fuel of 19th and 20th centuries fuel, currently losing its predominant role.

The perception of crude oil share by Polish students is of a different nature than the perception of coal. Percentage perceptions of crude oil share have a small scattering and the distributions for Poland and the EU do not differ clearly. Mean values of the share are underrated for both testes areas – for Poland by a few percent, for the EU – by several hundredths. The significant part of Polish students have no consciousness as to what large role crude oil plays in Polish economy and as far as the EU is concerned, one can be shocked by the lack of consciousness.

The perception of natural gas is totally different for the perception of coal and crude oil. A rather small scattering of perceptions of gas share appears here with a very large similarity of distributions for Poland and the EU. On the base of he analysis results it can be stated that half of the respondents overrated the level of this sector in Poland at the same time lowering the level of the EU gas sector. It should be stressed here that natural gas has the highest percentage of correct answers of all the sources of primary energy described so far. It may be due to the problem of gas deliveries as described by the mass media.

The results of the respondents' consciousness of nuclear energy in fulfilling the energy needs of Poland and the EU it was stated that only 58% of the students gave the correct answer – that Poland does not uses nuclear energy; the rest of them overrated the share of this source of energy for our country by giving shares other than zero. The share of nuclear energy in the EU energy industry was perceived rather correctly. Also this time one can speak about assimilation – a large percenttage of the respondents assumed that Poland does don differ from the EU in this respect. This fact contradicts the images that Polish society is against building nuclear plants in Poland. We can state that the society has been conscious of the fact that nuclear plants operate in many European countries for a long time.

The answers of Polish students to the question abort the share of renewable sources in energy consumption was even more intuitive than in former cases, since the term "renewable energy" comprises various forms of energy, which were not detailed during the survey. The perception of renewable sources, shaped by mass media rather than personal experience, is totally different from the perception of traditional energy sources. As the only source of energy, renewable sources were perceived correctly for Poland.

2.2. *Comparing the consciousness of Polish, Finnish and Italian students*

The comparative analysis covered mean values of perceptions for three samples (Polish, Italian and Finnish). Results of the perception of own country energy sector are shown in Figure 1.

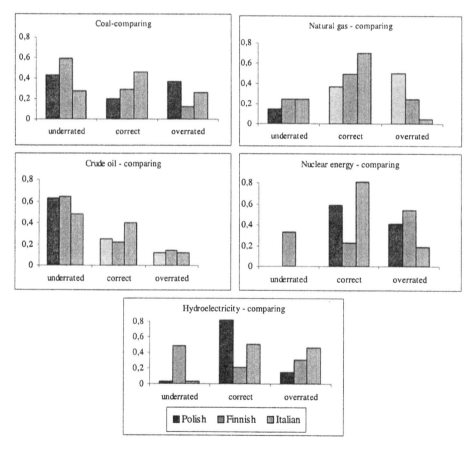

Figure 1. Comparison of shares perception correctness
of own country by Polish, Finnish and Italian respondents

The analysis of share perceptions of own country allows the conclusion that:
- the coal share in the own country was evaluated by the Polish and Italians correctly while the Finnish underrated it by a few percent;
- respondents of the three countries underrated the consumption of crude oil by their own economy by 7%;
- the share of natural gas in own economy is perceived by the Finnish correctly, the Poles tend to overrate and the Italians – to underrate it;
- perception of nuclear energy share is very interesting: perception for Polish and Italian students is totally different from the ones of their Finnish colleagues. Some Poles (42%) and Italians (18%) wrote in the questionnaire that their country uses nuclear energy, therefore mean share perceptions are slightly higher than zero. Finnish students, obviously under the influence of the educational campaign, overrated nuclear energy share in their own country;
- both Polish and Finnish students perceived renewable energy share correctly, whereas Italian students doubled their mean.

The synthesis of EU energy sector perception by the Finnish and Italian students as compared to the perceptions of the Polish respondents is shown in Figure 2.

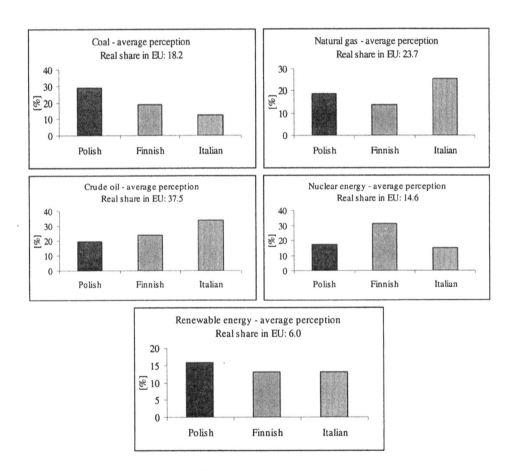

Figure 2. Comparison of shares perception correctness
by Polish, Finnish and Italian respondents for European Union

- When analyzing mean perceptions for EU we can say that:
- coal share in the EU was perceived correctly only by the Finnish students, their Polish colleagues overrated it highly and the Italian ones underrated it – the influence of their own country
  situation (assimilation);
- crude oil share in the EU was perceived by all the participants by similarity to their own country, therefore only the Italians have the correct mean (the Finnish have an underrated one and
  the Polish a very much underrated one);
- perception of gas share in the EU is identical to the one of crude oil, the only difference is that
  the Finnish students underrated the share;
- the Italians and Poles perceive nuclear energy in the EU correctly. This time, the Finnish complied to assimilation and perceive nuclear energy share in the EU at the target level of their country, thus over-perceiving the reality twice, this result shows the strength with which governments
  can influence public opinion;
- when perceiving the share of renewable energy, which in virtually every country is sponsored
  by the government, rather strange numbers appeared; although the Polish and Finnish students
  perceived medium shares of this source in their countries correctly, but in the remaining cases
  we see perceptions twice as much as the real shares, including all the perceptions for the EU.

2.2. Perception of dominant sources of energy

Finnish, Italian and Polish students answered the second part of the questionnaire about dominant
sources of energy in seven countries. The comparison of the answers of students of three nationali-
ties is shown in Table 4.

Table 4. Results of questions regarding the dominant source of energy

Country	Dominant source of energy	Percentage of students who showed the dominant source of energy correctly			Source of energy, which was most often indicated as dominant (% of indications)		
		Poles	Finns	Italians	Poles	Finns	Italians
France	Nuclear	43.6	59.4	39.7	Nuclear	Nuclear	Crude oil (45.4)
Norway	Renewable	61.8	34.9	18.4	Renewable	Crude oil (52.8)	Natural gas (36.2)
Germany	Crude oil	18.2	14.2	66.0	Coal (28.0)	Nuclear (47.2)	Crude oil
Hungary	Natural gas	21.5	20.8	31.9	Coal (52.6)	Coal (42.5)	Crude oil (39.7)
Poland	Coal	92.6	48.1	33.3	Coal	Coal	Crude oil (36.2)
Ukraine	Natural gas	26.9	18.9	53.2	Coal (42.2)	Nuclear (34.9)	Natural gas
Russia	Natural gas	48.0	18.9	53.9	Natural gas	Crude oil (42.5)	Natural gas

The results presented in Table 4 once again confirm a strong assimilation effect about other countries. Polish students think that just like in their country, coal is the dominant fuel, too. The perception of the Finns and the Italians was also shaped under the influence of their own country. The former attributed the dominant role of nuclear energy to as many as three countries and, since they do
not have a developed gas sector, they did not notice natural gas in the three countries. The Italians,
on the other hand, attributed the dominant role of oil and gas to all seven countries they had been asked about, forgetting about coal and nuclear energy, which are not important sources in their own
countries.

CONCLUSIONS

When analyzing the results of the survey it can say that the Poles more or less are aware of the role coal plays in Polish economy but they have virtually no idea about the importance of natural gas, crude oil and its by-products and nuclear energy. The Poles have very little knowledge about fuel consumption structure in the EU, the answers of the Poles show a strong assimilation effect, which manifests itself in levelling energy differences between Poland and the EU. All changes in home energy policy should be preceded by increasing the consciousness of the Polish society both for home and the EU energy sector. It is necessary to lead a constant, coordinated educational campaign, which will show the achievements of energy sectors of the EU and other European countries without bias.

The old-fashioned energy balance structure in Poland is due to low energy consciousness of the Polish society, which is unaware of the significant discrepancy between Polish energy culture and energy policy of the European Union and other countries of the continent.

At the end it should be stressed that educational campaigns have a feedback mechanism built in – the truth about energy shown to the public will not only facilitate appropriate decisions but also will make the public demand good decisions and finally, will make the public choose politicians suggesting modern energy solutions.

REFERENCES

[1] Attitudes Towards Energy 2006: Special Eurobarometer. European Commission. The European Opinion Research Group. http://europa.eu.int
[2] Centre of Public Opinion Examination (TNS OBOP) 2003: Poles about Miners, Coal and Restructuring Coal Mines, 1998–2003.
[3] Choudhury M.A. 1995: Ethics and Economics. A View from Ecological Economics. International Journal of Social Economics, Vol. 22, No. 2, pp. 40–60.
[4] Chukwuma C. 1996: Perspectives for a Sustainable Society. Environmental Management and Health, Vol. 7, No. 5, pp. 7–20.
[5] European Commission 2002: Energy: Issues, Options and Technologies. Science and Society. Eurobarometer.
[6] Ghobadian A., Viney H., James P., Liu J. 1995: The Influence of Environmental Issues in Strategic Analysis and Choice: a Review of Environmental Strategy among Top UK Corporations. Management Decision, Vol. 33, No. 10, pp. 46–58.
[7] Latek S. 2005: Nuclear Energy – the Majority is Pro! Energetyka, Vol. 60, No. 10, pp. 728–731.
[8] Łucki Z., Byrska-Rąpała A., Soliński B., Stach I. 2006: A Survey of Energy Consciousness of Polish Society. Polityka Energetyczna, Vol. 9, No. 2, pp. 5–63.
[9] Moshirian F. 1998: National Financial Policies. Global Environmental Damage and Missing International Institutions. International Journal of Social Economics, Vol. 25, No. 6/7/8, pp. 1255–1270.
[10] Pitarma R.A., Ramos J.E., Ferreira M.E., Carvalho M.G. 2004: Computational Fluid Dynamics. An Advanced Active Tool in Environmental Management and Education. Management of Environmental Quality, Vol. 15, No. 3, pp. 102–110.
[11] Pujari D., Wright G. 1996: Developing Environmentally Conscious Product Strategies: a Qualitative Study of Selected Companies in Germany and Britain. Marketing Intelligence and Planning, Vol. 14, No. 1, pp. 19–28.
[12] Schroeder J. 2002: Creating Energy Consciousness on the Campaign Trail. Clean Air – Cool Planet. http://www.cleanair-coolplanet.org
[13] Shaviv E. 1999: Integrating Energy Consciousness in the Design Process. Automation in Construction, Vol. 8, No. 4, pp. 463–472.
[14] Wolcott B. 2004: City Lights. Mechanical Engineering. Power & Energy, June 2004. http://www.memagazine.org

International Mining Forum 2008, Sobczyk & Kicki (eds) © 2008 Taylor & Francis Group, London, ISBN 978-0-415-46126-9

Capacity of an Abandoned Coal Mine Converted into High Pressure CO_2 Reservoir

Jan Palarski, Marcin Lutyński

Wydział Górnictwa i Geologii, Instytut Eksploatacji Złóż, Politechnika Śląska, Gliwice, Poland

ABSTRACT: In the article the concept of high pressure CO_2 storage in an abandoned coal mine is presented. Storage of CO_2 in abandoned coal mines is considered as one of the options of CO_2 sequestration. For the purpose of the study volume of mine voids in one of the polish coal mines was estimated. Storage capacity of a coal mine is the sum of CO_2 stored in mine voids, dissolved in mine-water and adsorbed on coal. The amount of CO_2 stored in mine voids and dissolved in water was calculated with the use of CO2-VR programme. Sorption capacity of a coal from a case-study mine was measured in a laboratory using volumetric adsorption apparatus. Experiments were performed at in-situ temperature of 22°C and 29°C. Results of the experiments were fitted into Langmuir adsorption model and Gibbs and absolute adsorption isotherms were plotted. Calculation of coal storage capacity was calculated based on absolute Langmuir constants and remaining coal reserves (mine losses). Storage capacity of a mine is mainly determined by CO_2 sorption on coal and was estimated to be 8 089 165 t. Out of the total amount of CO_2 stored, 56.5% (including ascertained potential) would be adsorbed on coal, 35.6% would be stored in mine voids and only 7.9% would be dissolved in water. The main factor influencing the storage capacity of a mine is not the sorption capacity of coal determined in laboratory but the remained reserves accessible for sorption. Storage capacity of a mine is also strongly limited by maximum allowable pressure of sealed shafts and overburden.

1. INTRODUCTION

Emission of gases such as carbon dioxide, methane, nitrous oxide and ozone into the atmosphere causes greenhouse effect and as a result global warming. Carbon dioxide is blamed to be the most responsible for the global warming and its emission is estimated to increase in the next 25 years by over 55%. The main source of carbon dioxide emission due to human activity is burning of fossil fuels which along with deforestation increase its concentration in the air. Primary energy production demand is projected to increase by over one-half between now and 2030 – an average annual rate of 1.6%. Fossil fuels account for 83% of the overall increase in energy demand in that period and will remain main source of energy. Due to that, carbon capture and storage (CO_2 sequestration) seems to be a viable way to reduce carbon emissions.

Among many considered options such as ocean storage or mineral trapping, geological sequestration is the most developed one and include storage in:
– deep saline aquifers,
– unmineable coal seams,
– depleted gas/oil reservoirs,
– mine caverns,
– underground mines.

Among the abovementioned sinks abandoned mines seem to be the least investigated. However, underground coal mines may be an interesting option for CO_2 storage sinks due to its abundance, usually close location to emission sources and relatively high storage capacity. On the other side, storage safety of trapped gas in abandoned mine is the main issue. Gas may leak through faults, cracks and fissures formed by former mining operation.

Carbon dioxide in abandoned coal mine will be stored as:
- free gas in mine voids, caving area or porous structures;
- dissolved gas in mine water;
- adsorbed gas in remaining coal seams.

Storage capacity of an underground coal mine is significantly increased by remaining coal seams where CO_2 is adsorbed [1], [2]. It is estimated that CO_2 adsorbed on coal may accounts for up to 80% of the total storage capacity.

The objective of the study is to asses the storage capacity of an abandoned coal mine converted into CO_2 storage site with focus on CO_2 adsorption on coal. As a case study one of the active Polish coal mines located in Silesia Coal Basin was selected and its storage capacity has been estimated based upon analysis of mine data and coal sorption experiments.

2. HIGH PRESSURE STORAGE IN ABANDONED COAL MINES

Underground coal mines were successfully converted into natural gas reservoirs in USA and Belgium. The oldest reservoir was located near Denver (Colorado, USA) and had been in operation since 1961 to 2003 supplying gas in peak demands for the city of Denver. The former subbituminous coal mine was extracting coal from two levels located at the depth of 210 m and 225 m. The cap rocks consisted of 20 m of claystones but the leak-off pressure of reservoir was determined by shafts which were able to withstand the pressure of 1.8 MPa i.e. about 75% of the hydrostatic pressure. Total storage capacity of a mine was estimated to be 74 mln m^3 of natural gas.

Two other reservoirs were located in the gassy Hainaut coalfield in southern Belgium (mines Anderlus and Peronnes). Both were characterized by a very low reservoir pressure. Publicly released data are available for Anderlus mine where workings were located between 660 m and 1100 m depth. However, some connections with flooded shallow workings exist and are locally only 10 m thick. The base of the overburden consists of clay, marl and chalk. Therefore, the reservoir pressure was limited to 0.35 MPa. At the low reservoir pressure adsorption of gas on coal plays a significant role. It was estimated that in Anderlus mine about 160 mln m^3 of gas is adsorbed on coal whereas 20 mln m^3 is a free gas in mine voids.

Currently, in Poland converting "Nowa Ruda" mine into underground natural gas storage site is being taken into consideration. The project estimates that the total storage capacity would be approximately 63 mln m^3 under the pressure of 0.25 MPa or approximately 153 mln m^3 under the pressure of 0.6 MPa. Out of the total amount of stored gas over 50% would be adsorbed on coal.

The main difference between temporary gas reservoir in underground coal mine and the CO_2 sequestration project is that in temporary gas reservoir infiltration water may be pumped through the time of storage operation. If we consider CO_2 sequestration in abandoned coal mine that option is economically (cost of water pumping) and technically unjustified because of the fact that the purpose of CO_2 sequestration projects is to store the gas for hundreds or even thousands years.

The concept proposed by Piessens & Dussar considers storing CO_2 under the pressure which is equal or exceeds hydrostatic pressure in the mine and prevents the influx of water. The phases representing the evolution of a mine reservoir, depicted in Figure 1, assume that at initial conditions free space is filled with CO_2 and mine is dry. The pressure of CO_2 at the initial conditions is equal or slightly higher than specific gravity of gas multiplied by the depth of mine. Because of the difference between density of CO_2 and water the pressure at the bottom of the reservoir will be lower than the hydrostatic in host rock. As a result, water will enter the reservoir (Fig. 1b). The rise of mine water will compress CO_2 and as a result pressure build-up will occur at the top seal of reservoir. Finally, reservoir pressure higher than hydrostatic will be reached at the top seal. As a result, CO_2 might migrate out of the reservoir into host rocks. This flux is neglected however, as its importance in non-permeable host rock is difficult to estimate and instead worst-case scenario (maximum pressure build-up at the top seal) is considered. The influx of formation water will continue until the pressure at each level of the reservoir filled with CO_2 is higher than hydrostatic pressure (Fig. 1c). The maximum pressure condition is not an equilibrium condition since the reservoir pressure is higher than the hydrostatic pressure in the host rock. Therefore, CO_2 will escape from the reservoir and migrate laterally into the host formation (Fig. 1d). This will result in a near complete flooding of the mine (Fig. 2e) but does not violate the terms for CO_2 sequestration as the gas will be trapped by the top seal.

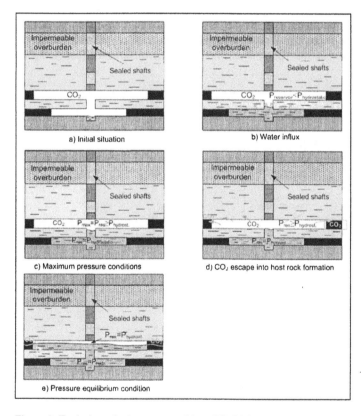

Figure 1. Evolution of mine converted into CO_2 high pressure storage site

As it was mentioned before, CO_2 in the mine will be stored as a free gas in mine voids, in solution in the mine water and adsorbed on coal. The volume of voids can be easily calculated based on mine data i.e. width, height and length of mine galleries, roadways, longwalls and porosity of gob area. The amount of CO_2 which will dissolve in water is a function of pressure, temperature and chemical composition (mainly salinity or Total Dissolved Solids). The amount of CO_2 adsorbed on coal can be estimated knowing the remaining reserves of coal and its sorption capacity. In order to estimate sorption capacity of coal laboratory experiments must be conducted.

3. LABORATORY EXPERIMENTS OF CARBON DIOXIDE ADSORPTION ON COAL

For the purpose of the study coal from the case study Polish mine "X" has been used. All measurements were done using single sample of coal obtained from a longwall face. The coal originated from a depth level of 650 m below the ground surface. Average temperature at that level was estimated to be 25°C and because it is planned to access deeper levels where temperature is higher it was decided to measure adsorption at the temperature of 29°C. The coal was crushed and grounded to a particle size of 0.149–0.425 mm. This exposed the internal surfaces of coal. Ground samples were placed in an environmental chamber to maintain 97% humidity and experimental temperature.

Figure 1 illustrates the schematic diagram of the experimental set-up consisting of a stainless steel sample cell, a two-way valve and high precision pressure transducer of maximum pressure 20.7 MPa, with a precision of 0.05% of the full scale valve. The coal sample under study is kept in a stainless-steel sample cell with a calibrated volume. A 2 μm filter is used to prevent coal particles from entering the valves. The total volume in the sample cell includes the volume occupied by the sample, voids space and adsorbed phase. The complete setup is placed in a temperature controlled water bath with a relative error < 0.2°C. Single gas adsorption experiment was performed at different temperature.

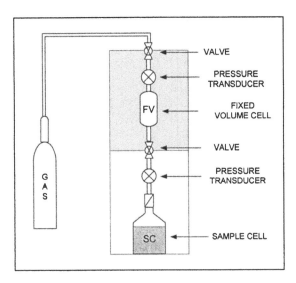

Figure 2. Schematic diagram of the volumetric adsorption apparatus used in a study

3.1. Adsorption calculation

The calculation is based on principle of volumetric measurement of gas sorption that states that adsorption removes the adsorbate gas molecules from the free phase to adsorbed phase, thus resulting in decrease in free gas pressure within the experimental system [4]. The number of moles of gas adsorbed (n_{sorbed}) during particular pressure step is the difference between the total number of moles of gas (n_{total}) introduced into the void volume of the sample container and the amount of free gas occupying the void volume (n_{free}) in the sample container.

$$n_{sorbed} = n_{total} - n_{free} \tag{1}$$

The total amount of gas fed into the sample container during a particular pressure step is calculated using the following equation:

$$n = \left(\frac{V}{RT} \cdot \left(\frac{P_1}{Z_1} - \frac{P_2}{Z_2} \right) \right) \qquad (2)$$

where P_1 and P_2 (MPa) refer to the pressure in the fixed volume container before and after the valve to the sample container is opened during a particular pressure step. V (cm^3) is the volume of the fixed volume container; Z_1 and Z_2 (-) are the compressibility factors of the adsorbate gas at the two pressures; R is the gas constant J/(mol · K) and T is the temperature (°C). The free gas amount is calculated using equation (2) above. However, in subsequent steps, free gas that is calculated also includes the free gas already present in the sample container from the previous steps.

The incremental free gas volume in the void space is calculated using the equation (3) below:

$$n_{free} = \left(\frac{PV_V}{RTZ} \right)_{current} - \left(\frac{PV_V}{RTZ} \right)_{last} \qquad (3)$$

where V_V is the void volume of the sample container. The values of compressibility factors at different pressures and temperature were calculated using NIST 14 software where Peng-Robinson equation of state is used.

3.2. Isotherm modelling

Experimental results have been fitted into Langmuir isotherm model which is commonly used in coal sorption modelling. The Langmuir isotherm is based on the concept of dynamic equilibrium between the rates of adsorption of gas on to the solid, and desorption from the solid surface. The equilibrium adsorption is achieved when the rate of molecules leaving the solid surface (evaporation) is equal to the sum of the rate of molecules condensed (adsorption) on the surface and molecules reflected from the solid.

The basic equation of Langmuir is shown below:

$$\frac{V}{V_L} = \frac{bP}{1 + bP} \qquad (4)$$

where V is the adsorbed volume (dm^3/kg) at equilibrium pressure P (MPa), b (-) is a constant known as the pressure constant, V_L is the maximum monolayer capacity, also known as Langmuir volume. At half coverage, that is, when sorbed volume is half of the Langmuir volume, the pressure value is referred to as the Langmuir pressure, or P_L. The above equation can be re-written in terms of V_L and P_L, the commonly used form of the Langmuir isotherm.

$$\frac{V}{V_L} = \frac{P}{P + P_L} \qquad (5)$$

where P_L is the inverse of b and other symbols have the same meaning.

In the Figure 3 Langmuir sorption isotherms for both temperatures fitted into experimental points are presented. Isotherms presented in Figure 1 are both Gibbs and absolute isotherms. Absolute adsorption takes into account volume of the adsorbed gas and was calculated using the following equation:

$$n_{ads} = \frac{n_{Gibbs}}{1 - \dfrac{\rho_{gas}}{\rho_{sorbed}}}$$

(6)

where n_{ads} and n_{Gibbs} are absolute and Gibbs adsorption in moles ρ_{gas} and ρ_{sorbed} are the CO_2 densities at gaseous and sorbed phase (g/ml). Density of adsorbed gas was assumed to be 1.18 g/ml which is the density of liquid CO_2 at the triple point [Arri et al. 1992].

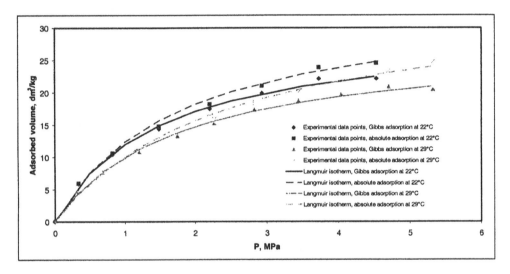

Figure 3. Experimental results of CO_2 adsorption on coal and fitted Langmuir isotherms

As it is seen in Figure 3 the amount of gas adsorbed decreases with temperature and it is a trend observed in many other experiments. It can be explained by changes in solid-gas sorption proper-ties at sorption sites with increasing temperatures.

The difference between Gibbs and absolute adsorption is higher at elevated pressures and rea-ches maximum of 20.8% and 13.7% at accordingly 22°C and 29°C. Calculated Langmuir constants, V_L and P_L, are compared in Table 1.

Table 1. Langmuir isotherm constants for CO_2 adsorption

Langmuir constants	Temperature of experiment			
	22°C		29°C	
	Gibbs	absolute	Gibbs	absolute
V_L, m³/t	29.9	34.6	28.0	35.3
P_L, MPa	1.5	1.9	1.8	2.5

4. CO_2 SEQUESTRATION CAPACITY OF AN ABANDONED COAL MINE

Total CO_2 storage capacity of a coal mine at the maximum pressure condition will be the sum of:

$$M_{CO2} = M_v + M_w + M_{ads} + M_a \qquad (7)$$

where: M_{CO2} – total mass of CO_2 storage stored in a mine, t, M_v – mass of CO_2 stored in mine voids as a free gas, t, M_w – mass of CO_2 dissolved in water, t, M_{ads} – mass of CO_2 adsorbed on remaining proven reserves, t, M_a – mass of CO_2 adsorbed on additional reserves (ascertained potential), t.

In order to calculate the first two parts of the equation i.e. mass of CO_2 stored in mine voids and dissolved in water the vertical reservoir simulator CO2-VR was used [3]. The mass of CO_2 adsorbed on remaining proven reserves and additional reserves was calculated on the basis of mine data and sorption laboratory experiments.

The vertical reservoir simulator CO2-VR uses the assumption of high pressure CO_2 storage in an abandoned coal mine presented in paragraph 2. Mine which was used for a case study currently has two extraction levels: 500 and 650 where volume of voids can be determined upon the length and cross-section of workings. In future, it is planned to access deeper seams located at 750 and 850 m depth. The volume of future voids can be roughly estimated knowing the amount of developed reserves and extraction ratios.

Current volume of mine voids i.e. workings was decreased by the consolidation factor of 0.5, gob area was assumed to have consolidation factors of 0.17, 0.13, 0.10 and 0.007 at the depths of 500, 650, 750 and 850 accordingly and was calculated using the equation:

$$V_g = R_e \cdot (\rho \cdot h)^{-1} \cdot h \cdot n[f(H)] \qquad (8)$$

where: V_g – void volume in gob area, m^3, R_e – mineable reserves, t, h – average seam thickness, m, ρ – coal density, t/m^3, n – gob consolidation factor in the function of depth.

Calculated vertical distribution of mine voids is presented in Figure 4 and was used as the input data.

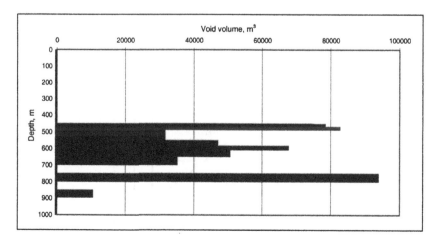

Figure 4. Void volume in a case study mine

One of the most important parameter in the CO2-VR programme is a maximum allowable pressure at the top seal given as a percentage of hydrostatic pressure. It was assumed that shafts are sealed up to the depth of 440 m and the CO_2 pressure cannot exceed 30% of the hydrostatic pressure at that depth i.e. 5.85 MPa. Other important input data used in programme are presented in Table 2.

Table 2. Main input data used in CO_2 storage capacity calculation by CO_2-VR

Input data	Value
Geothermal gradient, °C/100 m	2.8
Water density gradient, (kg·m³)/100 m	4.37
Water table depth, m	10
Total dissolved solids (TDS) in mine water, wt%	8.9
Total void volume accessible for CO_2, m³	17 481 616

Results of calculation are summarized in Table 3.

Table 3. Summary of CO2-VR calculation results

Result	Value
Total storage capacity of CO_2, t	3 500 000
Pressure at the top seal (440 m), MPa	5,43 (i.e. 124.7% hydrostatic pressure)
CO_2 dissolved in water, t	613 143
CO_2 stored in voids, t	2 886 857
Average CO_2 density, kg/m³	312

The calculation results are done for the pressure equilibrium condition which is defined as the state of reservoir at which the pressure at the contact of mine water and CO_2-fluid is equal to the hydrostatic pressure in the host rock. This means that the pressure in the reservoir filled only with CO_2 are higher or equal to the hydrostatic pressure. The depth of mine flooding at pressure equilibrium condition and CO_2 pressure compared with hydrostatic gradient are presented in Figure 5.

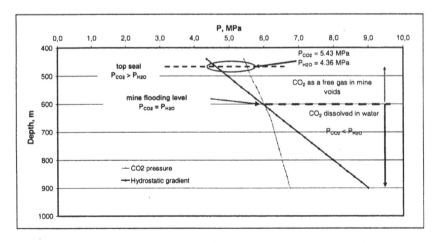

Figure 5. CO_2 pressure and hydrostatic gradient in a reservoir (abandoned mine)

The mine will be flooded up to the depth of approximately 620 m, above that level CO_2 will be compressed. Below the depth of 620 m CO_2 will be dissolved in mine water. Thus, the volume of voids accessible for "free" gas will be 48% of the total void volume i.e. 8.3 mln m^3. The calculated amount of gas that could be stored in a mine is only the amount of gas stored in voids and dissolved in water, not adsorbed on coal. In order to estimate the amount of CO_2 that could be adsorbed on remaining coal a simple approach based on the difference between the developed reserves and mineable reserves (i.e. mine losses) was used.

For the purpose of the study the following equation used to calculate the amount of CO_2 adsorbed on coal was developed:

$$M_{uds} = \left[m_1 - \left(\frac{(a+m)}{100} \cdot m_1 \right) \right] \cdot \left[\frac{V_L \cdot P}{P_L + P} \right] \cdot \rho_{CO_2} \tag{9}$$

where: M_{uds} – amount of CO_2 adsorbed on coal, t, m_1 – mine losses, t, $m_1 = (R_d - R_d \cdot k)$, R_d – developed reserves, t, k – extraction ratio, -, a – average ash content of coal, %, m – average moisture content of coal, %, P – pressure of CO_2 in reservoir at certain depth, MPa, V_L – Langmuir volume, m^3/t, P_L – Langmuir pressure, MPa, ρ_{CO2} – CO_2 density at normal conditions (0,001977 t/m^3).

Langmuir constants determined in laboratory as well as consolidation factors, mine developed reserves and recovery ratios based on accessible data from the mine company were used in calculation. Summary of data used in calculation are presented in Table 4.

Table 4. Summary of data used in calculation of adsorbed CO_2 on remaining coal

Level	R_d, th. t	k, -	m, %	a, %	V_L, m^3/t	P_L, MPa	P*, MPa
500	44 619	0,58			34.6	1.8	5.6
650	76 327	0,60	11.5	11			6.1
750	93 569	0,62			35.3	2.5	6.5
850	6349	0.65					6.7

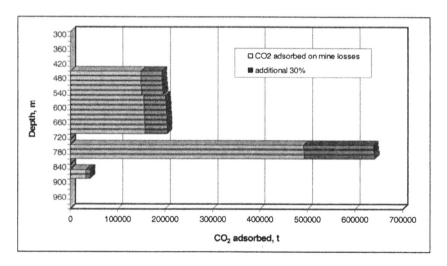

Figure 6. The amount of CO_2 adsorbed on mine losses with additional 30% of ascertained potential

The total amount of CO_2 adsorbed on remaining coal was estimated to be 3 530 127 t. Because of the fact that in calculation always "worst case" scenario was taken into account and the fact that there is a huge possibility of connections to remaining proven reserves this amount has been increased by 30% (ascertained potential). The extraction factor for proven reserves was estimated to be 0.33 which means that over 70% of coal remains in the deposit. In the figure 6 the amount of CO_2 adsorbed on mine losses increased with additional 30% is presented.

Finally, the total amount of CO_2 that could be sequestered in a mine which is the sum of all four components of the equation 7 was calculated to be 8 089 165 t.

CONCLUSIONS

Carbon dioxide sequestration in abandoned coal mines is one of the options of geological sequestration methods and so far has not been studied in detail. The main reasons for that are the safety features (possible leaks of gas through faults and fissures caused by former mine operation) and lack of precise data on remaining reserves and geological survey in abandoned mines. In order to consider such projects proper recognition of the storage site must be done before the mine closure.

Storage capacity of a mine is mainly determined by CO_2 sorption on coal. Out of the total 8 089 165 t that could be stored in mine, 56.5% (including ascertained potential) would be adsorbed on coal, 35.6% would be stored in mine voids and only 7.9% would be dissolved in water. Thus, the main factor influencing the storage capacity of a mine is not the sorption capacity of coal determined in laboratory but the remained reserves accessible for sorption. Storage capacity of a mine is also strongly limited by maximum allowable pressure of sealed shafts and overburden. Thus, with the high pressure of CO_2 in a mine the amount of accessible voids is increasing due to reduced flooding. Moreover, gas density is also higher contributing to increased storage capacity.

The calculated total storage capacity of a case study mine is not significant if compared to an average yearly coal fired power plant CO_2 emission. However, it is an option to reduce the emissions to a certain limits enforced by Kyoto protocol e.g. in European Union. In Europe, many abandoned mines exist and others are being closed. Most of them are closely located to power plants. Conventional coal fired power plants are able to increase their efficiency and emission only by a few percent and are not able to comply with limits unless brand new technology is developed. For example, the biggest Polish coal fired power plant "Belchatow" increasing the production by 3.1% has crossed the limit for CO_2 emissions by 1 480 949 t in 2005. In order to comply with the CO_2 emission limits the power plant would have to decrease the production or cross the limit by 1 800 000 t in 2009 and in the next two years by 5 600 000 t. Thus, the total amount of CO_2 which could be sequestered in an abandoned coal mine would be enough to comply with the limits for 3 or 4 years. If the CO_2 sequestration in abandoned coal mine would be combined with CH_4 recovery from remaining seams it could offset some of the investment costs.

REFERENCES

[1] Cisek W., Dybciak A., Landsberg W. 2001: Przebieg i doświadczenia z likwidacji kopalni "Nowa Ruda" w aspekcie przekształcenia wyrobisk dołowych na potrzeby podziemnego magazynu gazu. Przegląd Górniczy, Nr 7–8, Katowice 2001.
[2] Moerman, A. 1982: Gas Storage in Peronnes-Lez-Binch. S.A. Distrigaz Internal Report, 19 p.
[3] Piessens K., Dusar M. 2003: The Vertical Reservoir Simulator CO_2-VR. International Coal Bed Symposium Proceedings, the Paper No. 0347, Tuscalosa 2003.

International Mining Forum 2008, Sobczyk & Kicki (eds) © 2008 Taylor & Francis Group, London, ISBN 978-0-415-46126-9

Determination of the Size of the Fracture Zone Around a Road Working Based on the Results of Underground Convergence Measurements and Numerical Modelling

Stanisław Prusek

Central Mining Institute, Katowice, Poland

ABSTRACT: The publication presents the results of trials aiming at the determination of fracture zones around road workings exposed to the direct impact of the extraction pressure. The range of these zones was determined by means of the computer program Phase^2, which is based on the finite element method. The basis for the determination of rock mass destruction zones constituted the Hoek-Brown criterion for an elastic-plastic medium and the results of underground measurements of vertical and horizontal convergence as well as floor upheaval in roads.

KEYWORDS: Safety, mining, road support

1. INTRODUCTION

In the process of design and selection of road support for determined geological and mining conditions it is necessary to know the value of load acting on the support. In the case of road workings which are not exposed to direct impacts of the extraction pressure, the value of support load is calculated, determining the fracture zone (called also destruction zone or de-stressing zone) around the working and assuming that the rocks contained in the zone will statically influence the support. The calculation of the size and shape of the fracture zone is possible on the basis of numerous theories [1], [3], standards [10], or by means of numerical modelling [7], [8]. In underground conditions the determination of the fracture zone range around road workings enable endoscope tests, using cameras introduced into holes drilled around the workings [4], [5]. The assessment of the range and shape of the fracture zone is complicated in cases, when the road exists in the zone of an operating extraction front, as for instance bottom roads and top roads around longwalls. The extraction pressures can cause the growth of this zone along with the measuring of the longwall face [4].

This paper presents a trial with respect to the determination of the rock mass fracture zone around examined bottom and top roads around longwalls in the "Borynia" mine of the Jastrzębska Coal Company. In this mine within the period 2005–2006, in the framework of a targeted project, measurements of vertical and horizontal convergence in seven roads around caving longwalls were carried out. The main objective of these investigations was the identification of the deformation character of this group of bottom roads and top roads around longwalls in specific extraction conditions of the "Borynia" mine from the viewpoint of design of a new road support construction.

The determination of the rock mass destruction zone around the examined roads, on the basis of numerical modelling, consisted in the appropriate selection of model parameters that allow to obtain values of rock massive displacement, comparable with the values of measured convergence.

All numerical calculations were carried out by means of the program Phase^2 [6], [8], based on the finite element method. This program enables to model the rock mass in the form of a shield with unit thickness, being in the flat deformation state.

With regard to limitations relating to the volume of the present publication, the trial to determine the size of the fracture zone around roads has been presented by example of one longwall panel (longwall F-20), where convergence measurements in two roads along a longwall were conducted (F-20a and F-20b).

2. CHARACTERISTICS OF THE AREA OF CONDUCTED UNDERGROUND INVESTIGATIONS ALONG WITH THE DESCRIPTION OF THE MEASUREMENT METHODOLOGY AND RESULTS

The convergence measurements were conducted in the bottom and top roads F-20a and F-20b of the longwall F-20, in the top bed of the seam 405/1 (Fig. 1).

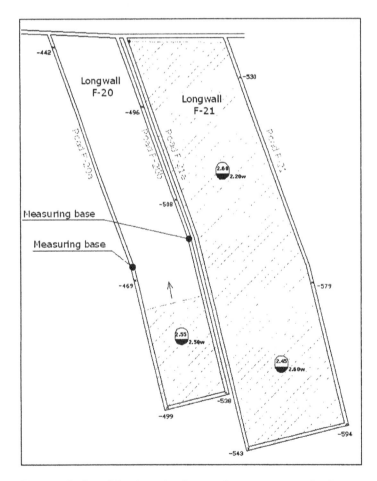

Figure 1. Outline of the place of underground measurement conducting, longwall panel F-20, 405/1 seam (top bed)

In the analysed extraction area the seam 405/1 occurs at the depth from 760 m to 810 m. The seam thickness in the top bed amounts to about 5 m, at the inclination of 10° to 17°. In the direct roof of the seam occurs a clunch layer about 4.5 m in thickness, and next arenaceous shale several metres in thickness. In the floor of the top bed of the seam 405/1 occur alternate thin clunch and carbonaceous shale layers; below them occur the seam bottom bed and a thick clunch layer with thickness reaching 20 m.

The longwall F-20, 154 m in length, was conducted with the height of about 2.6 m, with roof caving and average daily advance velocity reaching 5 m. The top and bottom roads around this longwall were liquidated beyond the extraction front. In the roads conventional arched yielding support (ŁP) was used, performed using sections of V32 type. The support frames, size 10, were built with the spacing of 0.8 m. In order to ensure space between support frames, both in the roof and on side walls, welded net was used.

For the determination of rock mass movements and support deformation occurring in the top and bottom roads F-20a and F-20b, on account of the impact of the longwall F-20 front a measurement method was used [9], applied since many years by the workers of the Central Mining Institute (GIG). In both roads, at the distances of 155 m and 170 m before the longwall F-20 front measuring bases were arranged, where a number of bench-marks were mounted in the roof, floor and side walls. Measuring cyclically the distances between individual measuring bench-marks, the values of vertical and horizontal convergence and floor rock upheaval with the nearing to the measuring bases of the longwall F-20 front were obtained. Additionally, the measurements of road support deformations through the determination of changes between characteristic points marked in a durable manner on the periphery of the support frame were carried out. All measurements were performed once a day, on each working day simultaneously with the measurement of measuring base distance from the longwall F-20 front. The results of rock mass measurements in the roads F-20a and F-20b were presented in Figures 2–3.

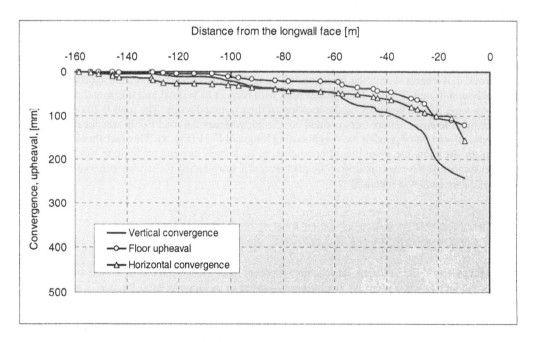

Figure 2. Average values of vertical and horizontal convergence and floor upheaval in the road F-20a

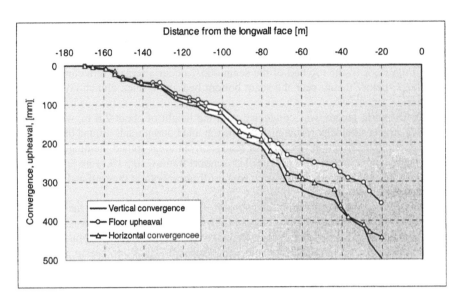

Figure 3. Average values of vertical and horizontal convergence and floor upheaval in the road F-20b

When analysing the measurement results we can observe differences in final values of rock mass movements that occurred in the roads F-20a and F-20b. Considering for comparison purposes the same distance before the longwall F-20 front, i.e. 20 m, in the first working the vertical convergence amounted to about 200 mm, at floor upheaval on the level of about 100 m. The value of horizontal convergence reached 110 mm (Fig. 2). These values should be considered as not high, taking into account the hitherto gained experience of the mine in the field of top and bottom road deformation. In the case of the second of examined roads (F-20b), the values of vertical and horizontal convergence and floor upheaval are several times higher than in the road F-20a. At the distance of 20 m before the longwall F-20 front, the vertical convergence amounted to about 500 m, with the share of floor rock movements within the limits of 360 mm. The horizontal convergence approached the value of about 440 mm (Fig. 3). Such differences with respect to vertical and horizontal rock mass movements were caused probably by the location of the examined roads. The road F-20a existed in a more favourable environment of two-sided coal solid, whereas the road F-20b was located in the one-sided environment of gobs of the neighbouring longwall F-21 (Fig. 1). This fact should be considered as the deciding one about the differentiation of rock mass movements in these workings, because no essential differences relating to the parameters of surrounding rocks were observed, and the support of these roads was identical.

3. DETERMINATION OF ROCK MASS DESTRUCTION ZONES AROUND TOP AND BOTTOM ROADS AROUND LONGWALLS

The determination of the rock mass destruction (fracture) zone around the examined top and bottom workings, on the basis of numerical modelling, consisted in the appropriate selection of model parameters, enabling to obtain values of rock massive displacements comparable with the values of convergence measured in underground conditions [8]. It seems that such approach can be determined as the analysis of inverse convergence of the road working, being a trial of presentation of changes occurring in the rock mass around roads on the basis of underground measurement results.

All numerical calculations were carried out by help of the program Phase[2] [6], which is based on the finite element method. This program enables rock mass modelling in the form of a shield with unit thickness, being in the flat deformation state.

The fracture zones around longwall top and bottom roads were calculated on the basis of the Hoek-Brown strength criterion [2]:

$$\sigma'_1 = \sigma'_3 + \sigma_{ci}\left(m_b\frac{\sigma'_3}{\sigma_{ci}} + s\right)^a$$

(1)

where: σ'_1 and σ'_3 – effective maximum and minimum stress at destruction [MPa], m_b – value of the Hoek-Brown constant for the rock massive, s and a – constant, determined on the basis of rock mass properties, σ_{ci} – uniaxial compressive strength of rock sample [MPa].

The parameters m_b, a and s are determined from the following relationship [2]:

$$m_b = m_i \cdot \exp\left(\frac{GSI - 100}{28 - 14D}\right)$$

(2)

where: m_i – constant for the undisturbed rock dependent on its type, determined on the basis of the three-axial compression test or on the ground of tabular data, GSI – parameter of rock mass quality (Geological Strength Index) determined for different geological conditions, D – destruction coefficient dependent on the type of rocks and extraction method.

In the case the value of the parameter $GSI > 25$, the remaining parameters of the Hoek-Brown criterion, i.e. s and a are determined from the following relationship [2]:

$$s = \exp\left(\frac{GSI - 100}{9 - 3D}\right)$$

(3)

$$a = \frac{1}{2} + \frac{1}{6}\left(e^{-GSI/15} - e^{-20/3}\right)$$

(4)

However, for the value of parameter $GSI < 25$ the constants s and a are determined from the relationship [2]:

$$s = 0 ;$$

$$a = 0.65 - GSI/200.$$

The strength coefficient according to the Hoek-Brown condition identified as the rock mass strength, is determined from the formula [2]:

$$W_{(H-B)} = \frac{\sigma_{max(H-B)}}{\sigma}$$

(5)

where:

$$\sigma = \sqrt{J_2} ;$$

$$\sigma_{mac(II-B)} = \frac{\sqrt{\left(1+\frac{tg\Theta}{\sqrt{3}}\right)^2\left(\frac{mR_c}{8}\right)^2 + \left(\frac{mR_cI_1}{12}+\frac{sR^2_c}{4}\right)} - \frac{mR_c}{8}\left(1+\frac{tg\Theta}{\sqrt{3}}\right)}{\cos\Theta} \qquad (6)$$

where: I_1, J_2, J_3 – invariants of strength tensor, MPa.

In the analysis of the state of stress and dislocation around a longwall top and bottom road work-ing an important role plays the modulus of longitudinal elasticity E. In order to determine the value of this parameter we can use empirical correlation relationships between the modulus E and other parameters such as: index of rock mass quality (GSI) or compressive strength.

Knowing the value of the compressive strength of rocks σ_{ci} we determine the modulus of longi-tudinal elasticity from the formulae [2]:

–   for $\sigma_{ci} < 100$ MPa:

$$E_m = \left(1-\frac{D}{2}\right)\sqrt{\frac{\sigma_{ci}}{100}}\cdot 10^{\frac{GSI-10}{40}}, \text{[GPa]} \qquad (7)$$

–   for $\sigma_{ci} > 100$ MPa:

$$E_m = \left(1-\frac{D}{2}\right)\cdot 10^{\frac{GSI-10}{40}}, \text{[GPa]} \qquad (8)$$

Because of the time consumption of numerical modelling, the calculations illustrating the im-pact of the advancing extraction front of the rock mass destruction zone around longwall top and bottom road workings were carried out with respect to cases, when the longwall front was at the di-stance of: 100 m, 50 m and 25 m, respectively, from the measuring base, in each of the examined workings.

Basing on the results of underground measurements of vertical and horizontal convergence (Fig. 2 and 3), calculation models were carried out aiming at the comparability of the value of obtained in models dislocations of rock layers with the values measured in underground conditions.

In all calculation models it has been assumed that the rock mass will be an elastic-plastic and isotopic medium. Moreover, the same boundary conditions were adopted, i.e. zero dislocations on all shield edges in the vertical and horizontal direction with regard to initial stresses resulting from the depth and average weight by volume of the overburden.

Because of the lack of possibility to model the support ŁP manufactured of the section of type V in the program Phase^2, for computational objectives such support was adopted in the form of beam elements with steel properties.

Additionally, in order to unify the calculation model it has been assumed that the longwall ex-traction front will move in the direction from beyond the shield to the shield front.

Numerical calculations illustrating the deformation size and rock mass destruction zone in the area of the roads F-20a and F-20b were carried out on a model performed in the form of a shield with dimensions 80×80 m. In the model it has been assumed that roads with dimensions correspond-ing with the support ŁP10 are located at depths 745 m and 785 m, respectively.

On the basis of obtained geological data it has been adopted that the side walls of roads consti-tute the coal of seam 405/1 (top bed). Directly above the roads occurs clunch 4.5 m in thickness; above the clunch occur in turn: arenaceous shale (16.0 m), coal seam 404/2 (bottom bed) (0.9 m), arenaceous shale (10.0 m) and clunch (8.0 m).

The direct floor in the area of roads is formed as a thin layer of clunch (0.1 m) and carbonaceous shale (0.9 m), below occurs clunch (1.5 m) and the bottom bed of the seam 401/1 0.9 m in thickness, and next a thick layer of clunch (20.0 m) and arenaceous shale (16.0 m). It has been adopted that all layers are inclined at an angle of 14°. The models of rock mass around the roads F-20a and F-20b were presented in Figures 4 and 5.

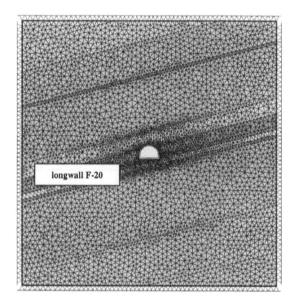

Figure 4. Rock mass model adopted for calculations in the area of road F-20a, seam 405/1 (top bed)

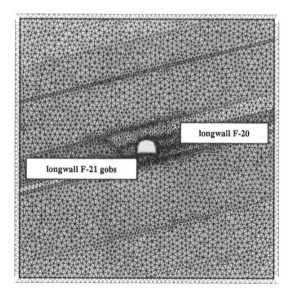

Figure 5. Rock mass model adopted for calculations in the area of road F-20b, seam 405/1 (top bed)

The basic strength-deformation properties of rock layers forming the models were presented in Table 1. These parameters were adopted on the basis of obtained results relating to penetrometric tests carried out in the roads F-20a and F-20b, on the basis of calculation results by means of the modulus RockLab [6]. The parameters of caving gobs were adopted on the basis of [11].

Table 1. Layer propertied in models of rock mass around the roads F-20a and F-20b

Type of rock	Young's modulus E [MPa]	Poisson's coefficient v	Compressive strength R_c [MPa]	Parameter of the Hoek-Brown criterion m	Parameter of the Hoek-Brown criterion s
Coal	1450	0.30	8.88	0.660	0.0004
Carbonaceous shale	1865	0.28	12.5	0.786	0.0010
Clunch	2432	0.26	26.4	0.985	0.0031
Arenaceous shale	4025	0.24	39.0	1.218	0.0042
Gobs	600	0.40	4.00	0.240	0.0001

Furthermore, it has been adopted in the built models that the road F-20a is located in both-sided solid coal environment, whereas the road F-20b is located in an one-sided gob environment on account of the earlier run of the caving longwall F-21 (Fig. 1). Between the gobs and the road F-20b is a coal pillar about 5 m in width.

Because of limitations concerning the volume of the publication, below in Figures 6–11 the results of calculations in the form of maps of rock mass strength, total displacements along with changes of the cross-section of roads F-20a and F-20b at a distance of 25 m before the longwall F-20 front were presented.

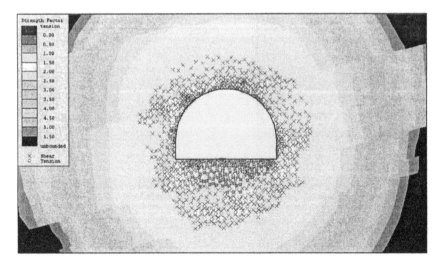

Figure 6. Map of strength according to the Hoek-Brown criterion along with the plasticization zone around the road F-20a – 25 m before the longwall F-20 front

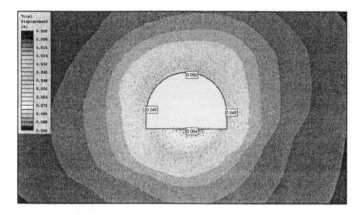

Figure 7. Map of total displacements along with maximum values
on the road F-20a outline – 25 m before the longwall F-20 front

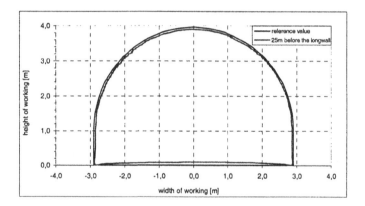

Figure 8. Change of the cross-section of road F-20a at the distance of 25 m before the longwall F-20 front

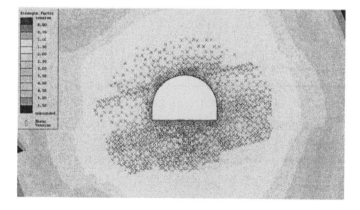

Figure 9. Map of strength according to the Hoek-Brown criterion along
with the plasticization zone around the road F-20b – 25 m before the longwall F-20 front

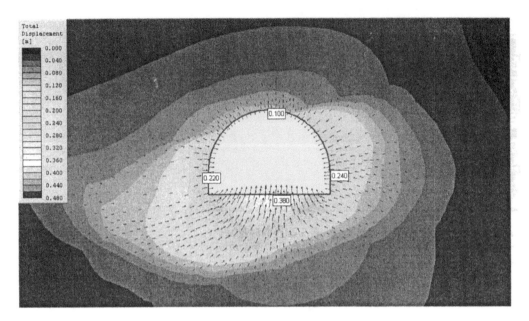

Figure 10. Map of total displacements along with maximum values
on the road F-20b outline – 25 m before the longwall F-20 front

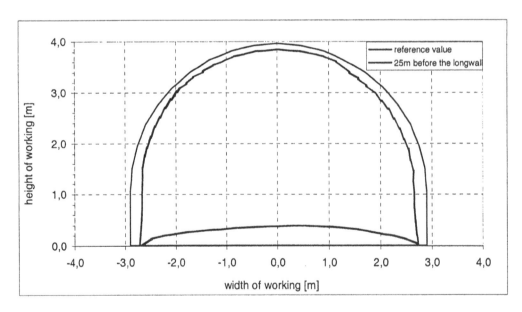

Figure 11. Change of the cross-section of road F-20b at the distance of 25 m before the longwall F-20 front

On the basis of numerical calculation results in Figures 12 and 13 the values of the maximum range of destruction zones in the roof, floor, and side walls of examined roads with respect to exceeding of the shear strength and tensile strength of layers were presented.

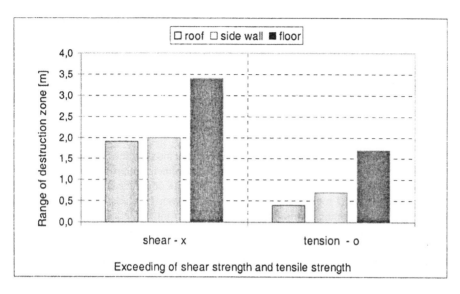

Figure 12. Destruction zone range around the road F-20a – 25 m before the longwall F-20 front.
Range of destruction zone [m]

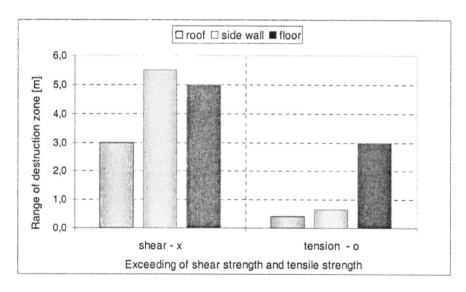

Figure 13. Destruction zone range around the road F-20b – 25 m before the longwall F-20 front

As a result of numerical modelling the shape and range of rock mass destruction zones at different distances from the operating extraction front were obtained. This publication presents the calculation results in the case of assumed distance from the longwall equal to 25 m. At this distance
from the operating longwall F-20 front the maximum values of the destruction zone range of rock
mass layers in the roof, floor, and side walls of the roads F-20a and F-20b were given (Fig. 12 and
13). When analysing the obtained results of numerical calculations we observe that the destruction

zone range of the rock mass around the road F-20b is higher in comparison with the road F-20a. In both cases dominate destructions of rock layers caused by shear strength exceeding. Moreover, in the analysed workings we can observe a higher destruction range of floor layers and side walls in relation to roof rocks, what is generally consistent with the obtained underground measurement results with respect to the convergence of these workings (Figs 2 and 3).

SUMMARY

The presented results of underground measurements of rock mass movements that took place in two roads around a caving longwall are a fragment of a wide measurement action conducted in 2005 and 2006 in several longwall panels of the "Borynia" mine. The main objective of these investtigations was the exact identification of rock mass behaviour in this group of mine workings, in very specific, difficult geological and mining conditions of the "Borynia" mine. Underground investigations were conducted in the framework of realisation of a targeted project; its subject was the design of a new construction of road support. Possessing wide research material, a trial was undertaken with respect to the determination of destruction zones of rock mass layers around examined top and bottom roads around longwalls, exposed to the direct impact of extraction pressure. Basing on underground measurement results of the vertical and horizontal convergence of roads and using the program Phase^2, which allows to carry out numerical modelling of rock mass behaviour, a probable image of the shape and range of destruction zones around workings at a determined distance from the longwall face was obtained. On the basis of numerical calculation results the maximum ranges of destruction zones in the environment of workings that occurred on account of exceeding of the shear strength and tensile strength in the roof, floor and side walls were determined.

The trial carried out to determine the range and shape of rock destruction zones around top and bottom roads around longwalls on the basis of underground measurement effects of rock mass movements gave promising results that can find application in the processes of selection and optimization of support of these workings.

REFERENCES

[1]   Borecki M., Chudek M. 1972: Rock Mechanics (in Polish). Publishing House "Śląsk", Katowice 1972.
[2]   Hoek E.: Practical Rock Engineering. Rocscience Inc., 1998, (www.rocscience.com).
[3]   Kłeczek Z.: Mining Geomechanics (in Polish). Śląskie Wydawnictwo Techniczne, Katowice 1994.
[4]   Majcherczyk T., Małkowski P.: Impact of the Longwall Front on the Size of the Fracture Zone Around a Roadway Workings of a Longwall (in Polish). Wiadomości Górnicze, No. 1, 2003.
[5]   Majcherczyk T., Niedbalski Z.: Investigations into the Fracture Zone Range Around a Roadway Working (in Polish). Przegląd Górniczy, No. 2, 2004.
[6]   Phase^2, User's guide. Rocscience 1998.
[7]   Prusek S., Masny W.: Trial of Numerical Modelling of a Road Working Located in the One-Sided Neighbourhood of Gobs Behind the Operating Caving Extraction Front (in Polish). Wiadomości Górnicze, No. 6, 2007.
[8]   Prusek S., Masny W., Walentek A.: Numerical Modelling of Rock Mass Around a Roadway Working Exposed to the Impact of Extraction Pressure (in Polish). Górnictwo i Geoinżynieria, AGH Quarterly, Year 31, No. 3/1, pp. 475–483.
[9]   Prusek S.: Deformation Size of a Longwall Road Maintained Behind the Caving Longwall Front by Means of Protective Pack (in Polish). Przegląd Górniczy, No. 7–8, 2003.
[10] Polish Standard – PN-G-05020: Underground Roadway and Chamber Workings – Vaulted Support – Rules of Design and Static Calculations (in Polish). 1997.
[11] Tajduś A., Cała M. 1999: Determinations of Parameters of Roadway Working Support Based on Numerical Calculations (in Polish). Geotechnics in Mining and Special Building, AGH, Kraków 1999.

International Mining Forum 2008, Sobczyk & Kicki (eds) © 2008 Taylor & Francis Group, London, ISBN 978-0-415-46126-9

The Knothe Theory in the World Mining

Anton Sroka
Technical Mining University, Freiberg, Germany

The research conducted by professor Knothe, the result of which was his PhD thesis, was spurred by the necessity to address a very practical and to this day valid problem of influence underground mining has on surface structures. The post-war period in Poland was marked by a dynamic development of the mining industry. Due to the fact that urban development in the 19th and 20th centuries was, geographically, strictly linked to the location of works, conflicts between mining companies and local communities were commonplace. The utilitarian character of the work is clearly seen by its title: "The influence of underground mining on the surface in view of security of surface structures". The thesis was defended at the Technical University of Mining and Metallurgy in Krakow in May 1951. The solutions contained therein were described in detail in 1953 in the first edition of a new quarterly "Archiwum Górnictwa i Hutnictwa" (The Mining and Metallurgy Archives).

The foreword for the first volume of the journal, devoted entirely to the problem of calculating the parameters of a subsidence basin, was written by prof. Witold Budryk, who, among others, wrote:

"Although in the past twenty-odd years some researchers managed to establish some laws ruling surface deformation, these were not sufficient to allow for predicting the impact of prospective underground exploitation.

Due to the utmost importance and urgency of the problem, the author prompted a team of researchers from the Technical University of Mining and Metallurgy, comprising St. Knothe, J. Litwiniszyn and A. Sałustowicz, to undertake the task of working out theoretical bases for several issues in order to fill the gaps they contained".

And further on:

"St. Knothe, who was the first to present the function, used a geometrical method, introducing in his thesis the term of a so called „influence curve". Basing on the available survey data he proved that the curve followed a gaussian model. This observation turned out to be correct and yielded high relationships between theoretical predictions and actuals.

The second very important achievement of Knothe's was describing theoretically the relationship between the rate of mining a seam and the magnitude of deformations on the surface. The assumptions taken here also turned out to be correct and in line with actual measurements".

The volume contained five papers presenting solutions, which are today called theories, namely:
1. J. Litwiniszyn – Differential equation of rock mass movement;
2. St. Knothe – Equation of a profile of a fully developed subsidence basin;
3. A. Sałustowicz – Subsidence basin profile as a yield of a strata on an elastic substratum;
4. St. Knothe – Influence of time on development of a subsidence basin;
and
5. W. Budryk – Determining horizontal deformations of the surface.

This very first volume of "Archiwum Górnictwa i Hutnictwa" is undoubtedly one of the annals of the Polish mining scientific thought. For the solutions presented in it their authors were honoured a National Award in 1953. Professors Witold Budryk, Stanisław Knothe, Jerzy Litwiniszyn and Antoni Sałustowicz to this day remain the most outstanding representatives of the Polish school of predicting surface deformations caused by underground mining activities.

The solutions proved themselves impervious to the passage of time and are still valid. The Knothe theory very quickly became well-known throughout the world. Still in 1953, thanks to Professor Neubert, German translations of the Knothe's papers published in the first volume of "Archiwum Górnictwa i Hutnictwa" were printed in "Bergakademie", a journal published by the Mining Technical University in Freiberg.

In 1956 the papers were published in Russian, and in 1957 the theory was presented at the international congress in Leeds and published in a mining journal in Hungary. In 1958 four of the papers were printed in a Peking Mining Institute journal.

In 1958 the Knothe theory was printed by a French and an English mining journal.

The platform which allowed to broadly present the achievements of the Polish mining think-tank, whose members, apart from professor Knothe, were professors Litwiniszyn, Sałustowicz and Smolarski, was German Academy of Sciences' International Bureau for Rock Mechanics in Berlin. In the years 1959–1969, i.e. during the period of its activity, the Bureau was the world's centre for rock mechanics. Sessions, work groups and conferences attended by most prominent professors were held yearly. Bureau's publications are to this day a treasury of information and mining knowledge, as the methods and solutions found in them can be applied by the industry to deal with its most vital problems. The proceedings contain the papers presented as well as the discussions they provoked. They bear witness to the great respect the Polish scientists boasted for their approach to the problem of the influence of mining on rock mass.

The achievements of the Polish school, and prof. Knothe in particular, promptly assumed their rightful place in the world literature. The place, which they hold to this day. Among others, the books and textbooks written by Martos (1967), Kratzsch (1974), Karmis (1987), Dżegniuk, Pielok and Fenk (1987), Whittaker and Reddisch (1989), Peng (1992) and most recently Jiang, Preusse and Sroka (2006), published in Chinese, could be named.

The textbook compiled by prof. Kratzsch can be particularly regarded to be of a world standard. Since its first issue in 1974 until 2004 it was reprinted in Germany 3 more times. It was also translated and published in Russian (1978), English (1983) and in Chinese (1984).

In the author's opinion, the development of the Knothe theory from the moment of its conception till today is directly linked with the development of computing technology.

The first calculations of subsidence basins were done with the use of so called graphicons, i.e. graphical integration grids. Using the method for rectangular areas the values of subsidence were read from a table for standard co-ordinates. The introduction of electronic computing technology expanded the calculation potential almost infinitely.

The solutions proposed recently are follow the idea of breaking mining areas into elemental fields. The solution for one element enables taking into account the inclination of the strata above, anisotropy, the time and distribution of convergence in the nearest vicinity of the mined deposit. In this case, a general solution based on a stochastic environment model proposed by J. Litwiniszyn was used. It forms a solid theoretical foundation for the Knothe theory and shows further directions for the expansion of its computing potencial (among others: Sroka (1984), (1987), Hejmanowski et al (2001)).

The true meaning of prof. Knothe's works I began to appreciate only during my time as a professor at the Technical University in Clausthal (formerly: Clausthal Technical Mining University), then, currently, as a professor at the Technical Mining University in Freiberg and particularly as an employee of Deutsche Steinkohle AG (German Coal Ltd.) directly responsible for mine extraction plans in the sense of minimizing the influence of extraction on the structures on the surface and the mine's infrastructure, e.g. shafts and drives and minimizing the induced seismicity.

The Ruhrkohle method used in the German mining industry is, in the mathematical sense, identical to the Knothe theory. The only difference between them lies in different parameters of the gaussian model, i.e. in different definition of the angle of draw.

On numerous occasions, when involved in projects and during my travels abroad, I met with broad interest in the Knothe theory and in its widespread use. Apart from the European countries I observed this mainly in countries such as the USA, Canada, South Africa, Australia and China, i.e. countries boasting developed mining industries.

For the past 10 years the author attended the Ground Control in Mining conference organised annually by the West Virginia State University in Morgantown. The event is simultaneously an annual American miners' society conference. A visible trend can be observed that increasingly large portion of the proceedings is devoted to surface protection problems, with most of the solutions presented basing on the Knothe theory.

The Knothe theory found application not only in the coal mining industry, but is also used to calculate surface subsidence in the case of:
- mining of copper ore and salt deposits by room and pillar method;
- storage caverns located in salt rock mass;
- exploitation of oil and gas deposits;
- tunnel constructions;
and
- flooding of closed mines.

The new applications contribute greatly to increased interest in the achievements of the so called Polish school not only in Europe but particularly in the USA and Australia.

To summarise: the theory formulated by professor Knothe in 1951 in his PhD thesis was, in the field of assessing the influence of mining on the surface and rock mass, truly revolutionary.

The theory presents a one-of-its-kind and unparallel input of the Polish mining scientific thought into the theory and practice of world mining. For the past 50 years it has found wide-spread use in the mining industry and has formed a base for further research for three generations of scientists, of whom many devoted to the subject a substantial portion of their careers.

This beyond any doubt proves that the Knothe theory till this day is valid and answers the requirements and expectations set forth by mining scientists and practitioners.

For, to quote Immanuel Kant (German philosopher, 1724–1804): "There is nothing more practical than a good theory".

BIBLIOGRAPHY

[1] Dżegniuk B., Fenk J., Pielok J. 1987: Analyse und Prognose von Boden- und Gebirgsbewegungen im Flözbergbau. Freiberger Forschungshefte. A729-Bergbau und Geotechnik, Markscheidewesen.
[2] Ehrhardt W., Sauer A. 1961: Die Vorausberechnung von Senkung, Schieflage und Krümmung über dem Abbau in flacher Lagerung. Bergbauwissenschaften, Nr 8.
[3] Hejmanowski R., Dżegniuk B., Popiolek E., Sroka A. 2001: Prognozowanie deformacji górotworu i powierzchni terenu na bazie uogólnionej teorii Knothego dla złóż surowców stalych, cieklych i gazowych. Biblioteka Szkoly Eksploatacji Podziemnej, IGSMiE PAN, Kraków 2001.
[4] Jiang Y., Preusse A., Sroka A. 2006: Angewandte Bodenbewegungs- und Bergschadenkunde (in Chinese), VGE Verlag GmbH, Essen 2006.
[5] Jung A. 1960: Möglichkeiten der praktischen Anwendung der Vorausberechnungsverfahren nach Knothe und Akimow. Freiberger Forschungshefte, A 146.
[6] Karmis M. 1987: Prediction of Ground Movements due to Underground Mining in the Eastern United States Coalfields.
[7] Department of Mining and Minerals Engineering, Virginia Polytechnic Institute and State University, Blacksburg, Virginia, 24061.
[8] Khair W.A. 1994: Surface Subsidence Prediction and Prevention Methods. Szkola Eksploatacji Podziemnej 1994, Jastrzębie Zdrój.
[9] Knothe S. 1951: Wplyw podziemnej eksploatacji na powierzchnię z punktu widzenia zabezpieczenia polożonych na niej obiektów. Praca doktorska, AGH, Kraków 1951.
[10] Knothe S. 1953: Równanie profilu ostatecznie wyksztalconej niecki osiadania. Archiwum Górnictwa i Hutnictwa, Kwartalnik, Tom I, Zeszyt 1.

[11] Knothe S. 1953: Wpływ czasu na kształtowane się niecki osiadania. Archiwum Górnictwa i Hutnictwa, Kwartalnik, Tom I, Zeszyt 1.

[12] Knothe S. 1957: Observations of Surface Movements under Influence of Mining and Their Theoretical Interpretation. Proceeding of the European Congress on Ground Movements, University of Leeds, April 1957.

[13] Knothe S. 1984: Prognozowanie wpływów eksploatacji górniczej. Wyd. Śląsk, Katowice 1984.

[14] Kolmogoroff A. 1931: Über die analytischen Methoden in der Wahrscheinlichkeitsrechnung. Mathematische Annalen, Verlag von Julius Springer, Berlin 1931.

[15] Kratzsch, H. 1974: Bergschadenkunde. Springer-Verlag, Berlin, Heidelberg, New York.

[16] Kratzsch H. 2004: Bergschadenkunde. 4 Auflage (2004), Deutscher Markscheider Verein e.V., Bochum 2004.

[17] Martos F. 1967: Bányakártan. Tankönyvkiado, Budapest 1967.

[18] Peng S. 1992: Surface Subsidence Engineering. Published by Society for Mining, Metallurgy and Exploration, Inc. Littleton, Colorado, ISBN 0-87335-114-2.

[19] Sroka A. 1984: Abschätzung einiger zeitlicher Prozesse im Gebirge. Schriftenreihe: Lagerstättenerfassung und -darstellung, Bodenbewegungen und Bergschäden, Ingenieurvermessung. Montanuniversität, Leoben, Austria.

[20] Sroka A. 1987: Selected Problems in Predicting Influence of Mining – Induced Ground Subsidence and Rock Deformations. Proceedings of 5th International Symposium on Deformation Measurements, Fredericton, Canada, 1987.

[21] Sroka A. 1994: Wymagania stawiane metodom prognozowania wpływów eksploatacji górniczych na górotwór i powierzchnię z punktu widzenia praktyki górniczej. Szkoła Eksploatacji Podziemnej 1994, Jastrzębie Zdrój.

[22] Sroka A., Hejmanowski R. 1995: Nietypowe zastosowania teorii Knothego do obliczania deformacji powierzchni. III Dni Miernictwa Górniczego, Ochrony Terenów Górniczych, Ustroń-Zawodzie.

[23] Sroka A. 1999: Dynamika eksploatacji górniczej z punktu widzenia szkód górniczych. Studia, Rozprawy, Monografie, Nr 58, Wydawnictwo Instytutu Gospodarki Surowcami Mineralnymi i Energią PAN, Kraków 1999.

[24] Sroka A. 2005: Ein Beitrag zur Vorausberechnung der durch den Grubenwasseranstieg bedingten Hebungen. 5. Altbergbau – Kolloquium, Technische Universität Clausthal, 3–5 November 2005, VGE Verlag Glückauf GmbH, Essen 2005.

[25] Whittaker B.N., Reddisch D.J. 1989: Subsidence. Occurrence, Prediction and Control. Developments in Geotechnical Engineering. 56, Elsevier, Amsterdam–Oxford–New York –Tokyo.

[26] Praca zespołowa 1968: 10 Jahre IBG 1958–1968. Internationales Büro für Gebirgsmechanik bei der Deutschen Akademie der Wissenschaften zu Berlin, Akademie Verlag GmbH, Berlin.

International Mining Forum 2008, Sobczyk & Kicki (eds) © 2008 Taylor & Francis Group, London, ISBN 978-0-415-46126-9

A Mining Software System at "Bogdanka" S.A. Underground Coal Mine – an Important Implementation of an Integrated Mineral Deposit Management System in the Polish Mining Industry

Jerzy Kicki
AGH – University of Science and Technology,
Department of Underground Mining, Cracow, Poland
Polish Academy of Sciences, Mineral and Energy Economy Research Institute, Cracow, Poland

Artur Dyczko
Polish Academy of Sciences, Mineral and Energy Economy Research Institute, Cracow, Poland

ABSTRACT: The characteristic feature of a modern civilization is intensification of economic development and associated with it consumption of mineral resources. Fear as to their sufficiency is therefore quite understandable, particularly with respect to non-renewable resources. The more so that the growth experienced by industrial production in the 20th century grew more than ten times and in the period after World War II consumption of mineral resources reached a level equal to their total production during the whole past history of mankind. The fear accompanies humankind from the dawn of history and is a result of the role mineral resources play in the development of a civilization [1].

Considering the above, any projects focused on optimization of deposit extraction should be considered worthy of attention. One of them – in the opinion of the authors of this paper – is undoubtedly a project aimed at building a computer system to aid decision-making in the process of preparing a deposit for extraction. The paper characterizes the basic assumptions underlying the framework of such a system and describes the key parameters of the implementation project currently under way at "Bogdanka" S.A. Coal Mine – for a number of years consistently the best-performing underground hard coal mine in Poland – in which the authors are involved [2].

1. BASIC ASSUMPTIONS FOR A SYSTEM AIDING DECISIONS IN THE PROCESS OF PREPARING A DEPOSIT FOR EXTRACTION AT "BOGDANKA" S.A. COAL MINE

Yet again "Bogdanka" mine sets new directions of development for the whole of the mining industry in Poland. This time round the area involved is computer technology and its use for the processes of mine design and deposit management. May difficulty and complexity of the problem be proven by the fact that, as yet, no underground mine in Poland has managed to implement a uniform and integrated mineral deposit management system. To be sure, KGHM "Polska Miedź" S.A. has made several bids at purchasing and implementing such a system, but ultimately the copper company always withdrew from the plans at some or the other stage of the purchasing procedure [3].

The aim of the project which has been under way at "Bogdanka" mine since the beginning of February this year, apart from the implementation of the system itself, which comprises several software tools, is to develop a new mine design planning and production scheduling methodology that would be closely linked to information on deposit structure and quality. This would allow to optimize production planning and coal extraction management and in the result effectively influence deposit management at the mine.

The system under implementation comprises the following parts:
– a digital model of the deposit;
– a digital mine map;
– a digital production schedule.

1.1. *A digital model of the deposit – the geologic part of the system*

The part of the system dealing with creating and managing a digital model of the deposit was based on Minex Horizon Program and a geological database containing:
– lithostratigraphic information;
– quality parameters;
– hydrogeological data.

The data was ordered and verified and only after this initial verification it became the input material subjected to interpretation by the project team and "Bogdanka" mine personnel. At this stage of the project, the standard features of Minex Horizon were found very useful. They helped in data analysis, verification of erroneous inputs and provided for interpolation and extrapolation of any deposit structure or quality model, interpretation and correlation of any of the parameters stored in the database and modelling of discontinuities (e.g. faults) [5].

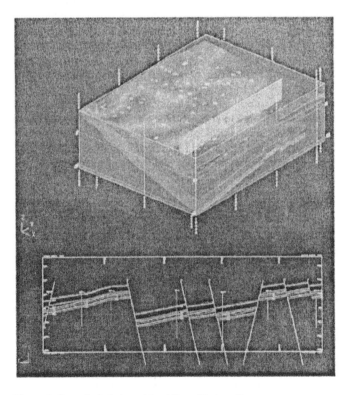

Figure 1. Deposit digital model – Minex Horizon Program

1.2. *Digital mine map – the survey part of the system*

In the course of the project at "Bogdanka" S.A. Coal Mine 70 maps drawn in 1:2000 scale were digitized with the use of Bentley's MicroStation software.

At the same time GMSI Modeller was utilized to create reports depicting the current state of the mine workings in 3D space and the actual coordinate system. Simultaneously with digitizing the maps an appropriate folder structure was created at the server, with user access duly restricted at file level. Appropriate procedures, regulating data sharing and archiving, were developed for the mine's survey department. They have already set in motion the procedure for obtaining a permit from the State Mining Authority to keep the base mine maps solely in digital format [4].

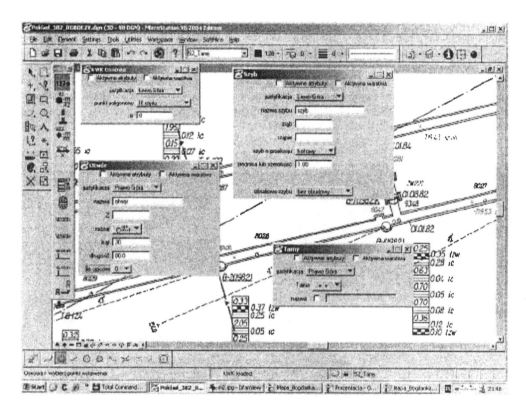

Figure 2. A digital map of "Bogdanka" S.A. Coal Mine underground workings created with Bentley's MicroStation

1.3. *Mine design and production scheduling – the technological part of the system*

Surpac Vision was chosen as the main program of the system dedicated to perform this task, as this software allows for comprehensive deposit management [5].

Supplementarily, in the technical and design part of the system, MineSched was used to cover the whole area of production planning and scheduling, encompassing the following activities:
- production planning (short and long term) and technical design;
- design of primary and secondary development and extraction workings;
- scheduling of the planned development and production;

- projecting quality parameters of the mined mineral based on geologic models;
- visualization of the planned workings and the schedule as an overlay over the geologic model;
- visualization of the whole mine design;
- scheduling material and equipment requirements for the planned workings.

Generally speaking, the design is done in 3D and it is based on a digital geologic model of the deposit describing both its structure and quality. The software provides for designing the workings in true dimensions, i.e. widths, heights, lengths and angle of dip of the digitally planned workings are the true dimensions of the respective workings underground [6].

The software allows the user to define any calendars of work and hence to create schedules for any periods of time. It could be a schedule covering one month divided into days, a one-year forecast providing monthly (weekly, daily) production figures, a 5-year plan, a 20-year plan or a life-of-mine plan. A very helpful feature of MineSched is its ability to automatically calculate volumes, tonnages and quality parameters of the mined mineral and waste rock for each defined scheduling period. The program allows to simulate extraction of a mining area by means of several faces in such a way that a certain pre-defined production figure is never exceeded. Each scheduling run ends by the program providing comprehensive reports detailing any of the default or user-defined parameters, such as coal tonnage, waste rock tonnage, coal quality parameters (sulphur and ash content, calorific value etc). The produced data can be presented in user-defined reports, arranged and filtered through, thus giving the user full flexibility in obtaining the required information.

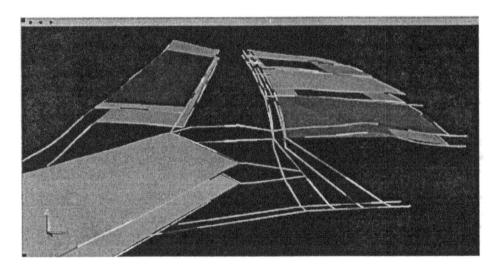

Figure 3. An example of graphical visualization of a development and production schedule.
Different colours symbolize different scheduling periods – Surpac Vision Program
assisted by MineSched software

2. BASIC GOALS OF THE PROJECT

In general, it must be said that the project was not intended to be confined to a mere implementation of software in the technological departments of the mine and accomplishing this will not render it completed. Right from the start its main objective was a total integration of work flow of all mining and mining-services departments and making full use of the computer system by obtaining the most optimal extraction scenario. To achieve this objective it was necessary to develop an optimal data-flow routine, which would allow the personnel of all technical departments of the mine, i.e. survey, geologic and planning, access relevant information depending on their qualifications and duties. The work conducted with this goal in mind have already resulted in creating a digital model of the deposit and a detailed plan of its extraction. The formats of the maps, sections, listings, graphs and schedules produced were compliant with the requirements of the Polish regulations, standards and internal company procedures, which in turn, are being adapted to the EU standards [7].

Achieving the planned objectives demanded from the project team and the mine technical personnel gathering and analysing an enormous amount of data, assays, measurements and other relevant information. All this work served to develop a multicriterion digital deposit development and extraction design, conforming with the requirements of sustainable development of extraction, assuring good environmental protection standards, providing for rational utilization of the mineral.

The optimal plan of development and extraction of the deposit at "Bogdanka" mine arrived at in this way is meant to form a base for further verification, necessitated by constant acquisition of new data by the mine technical and economic services [6].

This ongoing process will ultimately provide for:
- improvement in the area of mineral quality and production economics management;
- improvement in maintaining coal quality parameters at desired levels, better utilization of the deposit;
- improved safety of mining;
- increased efficiency, increased working times of technological units.
  Additionally, each user of the system will be able to:
- update data, according to his duties and authorisation level;
- analyse data presented in graphical and text (table) formats;
- edit data presented as standard and custom-created maps;
- have access to summaries and reports;
- connect to the system on-line.

The software system aiding deposit management, which is currently under implementation at "Bogdanka" mine, carries an enormous potential and involves processing a gigantic amount of information and data. The extent to which it benefits the mine economically depends entirely on its utilization by the mine personnel, who are currently subjected to an extensive training program to help them cope with the new challenge.

SUMMARY

The most important feature distinguishing the implementation project described in this paper from other implementations of this kind carried out by software companies in Poland and abroad, is the consistency in adhering to the accepted course of action and the principal mission of the project, namely an engineering approach to managing a mine, respecting the experience of the mine personnel and compliance with mining practices. The main strong points of our solution are openness and modular structure of the system, which integrates the existing technological solutions around one platform streamlining and optimizing activities of the mine [3].

The project team expects to see the first economic benefits of purchasing the state-of-the-art software tools already at the stage of designing the system.

This would come as a result of:
- integrating and simplifying the, broadly understood, mining processes involved in deposit extraction;
- implementing flexible multiscenario scheduling, short-, medium- and long-term;
- obtaining full control over quality and integrity of the gathered data;
- speedy verification of source data;

- on-line production data collection and its transfer to economic departments for use in developing integrated business plans;
- possibility for simultaneous use of the same data by many users;
- eliminating duplication of data storing and processing;
- implementation of central data protection mechanisms;
- limiting access to data to authorised personnel and departments;
- creating custom reports and maps for the purpose of mine management and outside institutions (ministries, mining authorities, municipalities);
- immunity of the deposit management system to quantity and complexity of the information describing the deposit;

All the benefits named above can, and probably will, directly influence future financial results of the company by:
- increased production (knowledge of the deposit, possibility to simulate multiple extraction scenarios, verification of scenarios for their economic viability, selecting an optimal scenario);
- optimising purchasing and stores procedures in the area of basic consumables and material, constructing day-to-day supplies plan;
- potential for speedy reaction to changing market situation;
- freeing resources in terms of qualified personnel from survey, geologic and mining departments, who can then be directed to research and development work, which would allow to obtain higher quality of knowledge of the deposit without incurring any additional costs.

The productive start of the system described above is planned for 15 October 2007. The project team are currently busy with configuration work and personnel training. Nonetheless, even at this early stage one may venture to draw a conclusion, whose essence may be condensed to a single sentence: to avoid being rejected by future users the structure of the system, in its detail, must be strictly adapted to the specific and unique character of the client company, in this case a coal mine, its organizational structure and its existing state of computerization.

REFERENCES

[1] Kicki J.: Wystarczalność surowców mineralnych a wystarczalność zasobów złóż – historia i aktualia. Wyd. IGSMiE PAN, Kraków 2004.
[2] Stopkowicz A., Kicki J., Dyczko A.: Niektóre aspekty zarządzania złożem na przykładzie złoża rud miedzi. Wyd. IGSMiE PAN, tom 23/2007, Zeszyt Specjalny nr 4, Kraków 2007.
[3] Dyczko A., Klos M.: System wspomagania decyzji w procesie przygotowania złoża do eksploatacji w kopalni węgla kamiennego "Bogdanka" S.A. – założenia, funkcjonalność i przepływ danych. Wyd. IGSMiE PAN, Materiały Konferencyjne, Szkoła Eksploatacji Podziemnej 2008, Kraków 2008.
[4] Kaczmarek A., Klos M.: Numeryczna mapa górnicza jako ważny element na przykładzie LW "Bogdanka" S.A. Wyd. IGSMiE PAN, Materiały Konferencyjne, Szkoła Eksploatacji Podziemnej 2008, Kraków 2008.
[5] Siata E.: Model geologiczny złoża i jego rola w zarządzaniu produkcją. Wyd. IGSMiE PAN, Materiały Konferencyjne, Szkoła Eksploatacji Podziemnej 2008, Kraków 2008.
[6] Praski M., Wachelka L.: Harmonogramowanie produkcji z wykorzystaniem narzędzi informatycznych na przykładzie LW "Bogdanka" S.A. Wyd. IGSMiE PAN, Materiały Konferencyjne, Szkoła Eksploatacji Podziemnej 2008, Kraków 2008.
[7] Kicki J., Dyczko A.: Górniczy System Informatyczny w podziemnej kopalni węgla kamiennego "Bogdanka" S.A – ważne wdrożenie zintegrowanego systemu wspomagania decyzji w procesie przygotowania złoża do eksploatacji w polskim górnictwie. Wyd. IGSMiE PAN, Materiały Konferencyjne, Szkoła Eksploatacji Podziemnej 2008, Kraków 2008.

International Mining Forum 2008, Sobczyk & Kicki (eds) © 2008 Taylor & Francis Group, London, ISBN 978-0-415-46126-9

Maintenance of a Mine Working Behind the Longwall Front Using Roof Bolting

Andrzej Nierobisz
Central Mining Institute, Katowice, Poland

ABSTRACT: The paper presents the way of maintenance of a mine working behind the longwall front using roof bolting. This technology was based on the use of roof bolts about 6.0 m in length with underslung straight sections of V-25 support in the axis of the excavation. In the work the way of design realisation, its implementation in practice and obtained measurement results were descri-bed.

KEYWORDS: Rock mass, longwall gate roads, support, bolting

1. INTRODUCTION

In the situation of necessity to maintain a road behind the longwall front, from the side of the cav-ing side wall abutments of different types were carried out, such as for example: rows of breaker props, wooden chocks, roadside packs made of plastics and others.

Investigations carried out at the Central Mining Institute [1] have shown that in consequence of longwall mining the rock mass in the longwall surroundings losses its physical continuity, main-taining only the geometric continuity. The rock mass layers pass to a set of elements with shape si-milar to parallelepipeds, connected to each other and mutually tightened by horizontal forces, ari-sen as a result of layers tendency to increase their volume after fracturing. The place of occurrence of first rock layer fractures in the roof changes its location in relation to the longwall face with the increase of its life (in terms of distance). Usually the first rock mass pressure symptoms are obser-ved about 50 m before the longwall front. The largest mining exploitation impacts are observed behind the longwall up to the distance of about 100 m, and then the deformations of support and excavation floor become stabilised.

Previously many different support solutions in order to maintain longwall gate roads subject to mining extraction impacts were used [3], [5]. During the last period of time successfully roof bolt-ing in order to maintain the stability of longwall gate roads was started [10], [16]. With regard to the facts mentioned above, in one of the mines trials were undertaken aiming at the maintenance of a longwall gate road by means of bolting.

The paper presents the methodology of roof bolting design, ways of its realisation and results of underground measurements relating to excavation stability.

2. DESCRIPTION OF THE GEOLOGICAL AND MINING CONDITIONS AND INVESTIGATION RESULTS

The thickness of the seam 407/1 in the discussed area ranged from 1.3 to 1.6 m. Its inclination fluc-tuated between 3 and 14 degrees, locally it reached 18 degrees. The discussed gallery j85 occurred at a depth of 650÷850 m. The roof rocks constituted: 3.0 m of mudstone turning into arenaceous shale, 4.5 m of fine-grained sandstone and 1.5 m of mudstone. In the floor occurred 0.6 m of shale with coal, 26.5 m of mudstone changeably sanded up.

The area of gallery j85 was counted among: the first degree of water hazard, class B of dust hazard, first category of methane hazard. The seam was classified among seams that are not subject to the rockburst hazard.

In consequence of underground and laboratory tests carried out the following geomechanical parameters of coal and roof rocks were obtained [11]:
- average compressive strength of roof rocks: 30.5 MPa;
- average compressive strength of coal from seam 407/1: 13.9 MPa;
- average water logging rate index: 0.80;
- average fissility index RQD: 77%.

The above mentioned results constituted the basis to carry out the roof bolting design.

3. ROOF BOLTING DESIGN

The design calculations of roof bolting parameters were carried out using two methods:
- analytical-empirical method;
- numerical modelling method.

The support parameters obtained by means of both of these methods were compared each other; more advantageous from the viewpoint of mine working stability to be used in practice were adopted.

3.1. Analytical-empirical method

3.1.1. Design methodology
The proceeding procedure comprised the following steps:
1. At first the state of stresses occurring in excavation side walls was analysed. The analysis takes into consideration the general concentration of forces, resulting from the location depth, and side wall rock strengths. These stresses are compared with the effective compressive strength of side wall rocks and on the basis of the Saint-Vernant criterion the fracture range in side walls is determined [7].
2. Next the calculations of the fracture zone range in the road working's roof were carried out in a situation, when it occurred in the one-sided gob surroundings (about 100 m behind the longwall front) [1]. The bolts should be fastened above the range of the above mentioned zone.
3. The pressure arch determined according to the model mentioned above served the calculation of the predicted load of road excavation support.
4. The load obtained is multiplied by coefficients taking into consideration the impact of overworking, underworking, edges, faults, neighbouring excavations and inclination.
5. The calculation load was compared with the total support frame resistance; next the distance between bolts in the excavation's axis was calculated.

3.1.2. Input data for design calculations
The input data serving the calculation of the state of stresses and rock mass effort comprise information on strength parameters of roof rocks and side wall rocks as well as data concerning the mining and geometric characteristics of the considered mine working together with the surrounding area. The strength parameters of roof rocks, constituting the calculation basis, were obtained on the basis of penetrometric and laboratory tests.

The fundamental data were as follows:
- average depth of excavation occurrence: 750 m;
- excavation dimensions in the breakout: 5.2×3.5 m;

- fissility coefficient: RQD = 77%;
- average penetrometric strength of roof rocks: 30.5 MPa;
- average coal strength: 13.9 MPa;
- average water-logging rate index: 0.8;
- minimum load-bearing capacity of individual bolt: 320 kN;
- distance between standing support rows: 0.75 m;
- type of standing support: ŁP9/V29/A;
- tremors: did not occur;
- former mining operations: the impact of seam 407/2 edge was assumed;
- abutment from the caving side: nine-point wooden chock;
- wooden chock resistance: 150 kN/m [13];
- load-bearing capacity of ŁP9/V29/A frame: 570 kN [12].

3.1.3. *Results of design calculations*

The analysis carried out relating to rock mass roof load has shown that it can amount to 1479 kN/m of excavation.

The predicted range of the rock breaking zone will be equal to (Fig. 1):
- in the roof: 4.5 m;
- in side walls: 2.7 m.

The calculation programme used to the calculation of roof bolting parameters for the given rock mass strength profile and assessment of predicted load gives the following calculation results:
- distance between bolts in the excavation axis: 0.8 m;
- length of bolts fastened in the rock mass: 5.5 m.

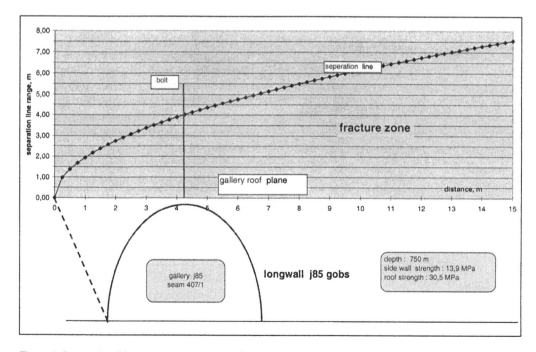

Figure 1. Prognosis of fracture zone range for gallery j85, seam 407/1 – panel behind longwall face

3.2. Numerical modelling method

3.2.1. Design methodology
Numerical modelling was carried out using the Finite-Difference Method Program FLAC v. 4.0. This program is designed to solve problems regarding geotechnology and mine or ground engineering and not only.

Every element behaves in conformity with the assumed linear or non-linear stress-deformation, settlement of equilibrium state and model response testing till the moment, when acceptable results have been obtained [4]. The support should also fulfil one fundamental condition, i.e. the bolt load should be lower than the load-bearing capacity of bolts.

3.2.2. Input data for numerical calculations
For calculation purposes has been assumed that the designed excavation occurs at the depth of 750 m. The situation was considered with installed standard yielding support, which was mounted by means of two pairs of steel bolts 2.2 m in length. Additionally, because of expected impacts of the caving zone, in the excavation axis a IR-46 bolt was modelled, 5.5 m in length, inserted in a section 8 m long.

The built model had dimensions 50×60 m and was divided into a calculation network with meshes 1.0×1.0 m. At side edges displacement conditions were adopted. At both side edges zero horizontal displacement and at the bottom edge zero vertical displacement were assumed. The lithological system of layers was adopted on the basis of underground tests. The top edge of the model was loaded using vertical pressure equal to 18.13 MPa in such a way that within the planned excavation existed pressure of about 18.75 MPa.

The medium's model adopted for calculations is the Coulomb-Mohr model. In calculations also the force of gravity was considered. The rock mass mechanical parameters adopted for calculations were presented in Table 1. In Tables 2 and 3 the basic bolt parameters, and in Table 4 the support ŁP parameters were presented [6], [8], [14], [15]. The model was presented in Figure 2.

Table 1. Mechanical parameters of rock mass

Material	Density	Internal friction angle	Coherence	Dilatation angle	Young's modulus	Poisson's ratio	Tensile strength
	ρ	Φ	c	ψ	E	N	Rr
	kg/m^3	degrees	MPa	degrees	GPa	–	MPa
Coal – 1	1400	20	1.6	5	2.0	0.25	0.8
Mudstone – 2	2700	25	3.0	7	7.0	0.23	1.2
Mudstone – 3	2700	25	3.3	7	7.2	0.23	1.4
Arenaceous shale – 4	2600	27	4.0	10	10	0.23	2.3
Sandstone – 5	2800	30	5.0	14	13	0.22	2.4
Gobs – 6	1200	20	0.1	3	0.7	0.4	0.1

Table 2. Mechanical parameters of roof bolts

Bolt length	Bolt diameter	Load-bearing capacity	Young's modulus of bolt	Binder rigidity	Binder coherence
L	Φ	–	E	Kbond	Sbond
M	m	MN	GPa	MN/m/m	MN/m
2.4	0.020	0.20	205	31.51	0.11
5.5	0.024	0.32	205	31.51	0.11

150

Table 3. Mechanical parameters of ŁP arch support

Cross-section field	Young's modulus	Moment of inertia	Steel density
A	E	J	Ro
m²	GPa	m⁴	kg/m³
0.0032	205	0.000027	7500

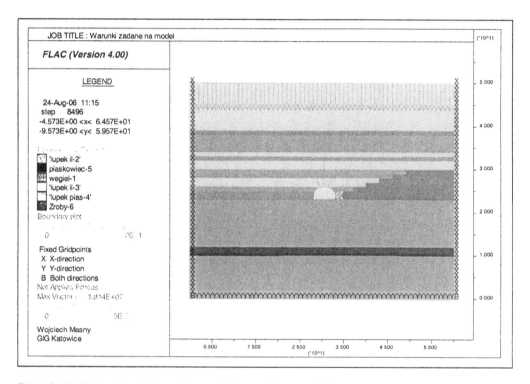

Figure 2. Model together with boundary conditions and force applied

3.2.3. Results of numerical calculations

In consequence of analyses carried out the following results were obtained:

− maximum axial forces occurring in steel bolds will amount to 52 kN. In the case of bolt load-bearing capacity amounting to 150 kN we can state on the basis of numerical analyses that the tensile strength of steel bolts will not be exceeded. In the case of IR-4C string bolt sectionally inserted the maximum axial forces should not exceed 114.2 kN, what according to the adopted assumptions allows to state that the load-bearing capacity of bolts was not exceeded. The axial forces in bolts were presented in Figure 3;

− maximum support displacements will follow from the side of gobs and will be equal to about 74 mm (Fig. 4);

− maximum displacements in the roof should not exceed 60 mm (Fig. 5). On account of the location of plasticizing zones, which in some approximation can be identified with destruction zones, floor upheavals are possible.

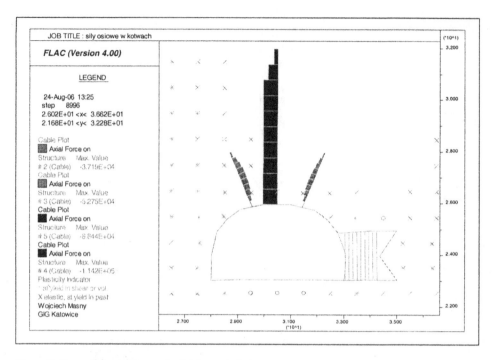

Figure 3. Zone of plasticizing and axial forces in bolts

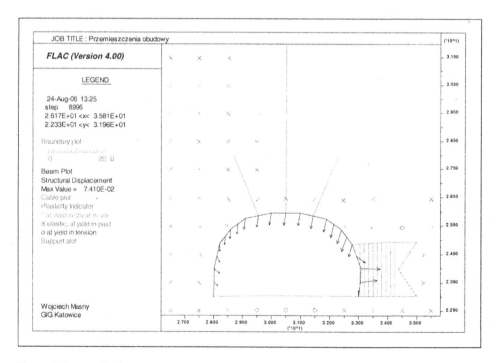

Figure 4. Support displacements

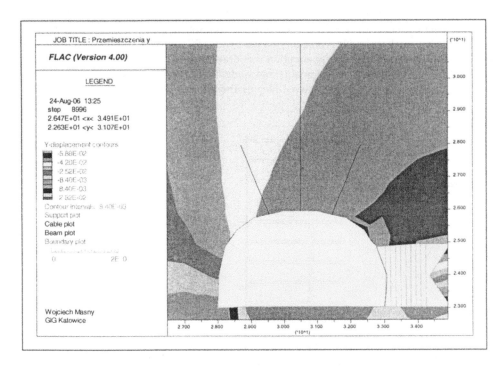

Figure 5. Values of vertical displacements

3.3. *Proposal of way of gallery j85 reinforcement*

In consequence of design calculations carried out using two independent methods, the realisation of gallery j85 reinforcements in conformity with the following principles was proposed:
- every roof-bar arch should be fixed using two pairs of rod bolts of many type distributed symmetrically at the distance of about 1.0 m from the excavation front;
- the minimum load-bearing capacity of the bolds mentioned above should not be lower than 150 kN;
- the total bolt length should not be lower than 2.4 m, each 0.6 m in length;
- along the excavation axis should be installed V+25 sections 2.5 m in length covering 3 frames;
- the sections should be fixed by help of bolts of IR+46 type with load-bearing capacity 320 kN, 6.0 m long;
- the IR-4C/320 bolts should be placed along the excavation axis at distances every 0.8 m;
- bolting should outstrip the longwall face (minimum 60 m);
- the side wall from the side of longwall j85 caving should be protected by means of a wooden chock or pack made of plastics with resistance (load-bearing capacity) not lower than 150 kN/m.

4. ROOF BOLTING CARRYING OUT

During the implementation of the discussed design one has stated that the use of roof-bar arch fixing by means of 4 bolts is hindered because of technical and operational reasons. In connection with the above one has renounced fixing, using instead of it from the side of the coal face a rail stringer supported by props (Fig. 6).

153

The reinforcement by means of bolts was carried out in conformity with the principles descrybed above. The fundamental abutment protecting the gallery's j85 stability was a nine-point wooden chock. During its execution particular attention was drawn to the solidity of its performance consisting in the selection of suitable wood quality and wedging under the excavation roof in order that the chock could as soon as possible take over the pressure from the caving side.

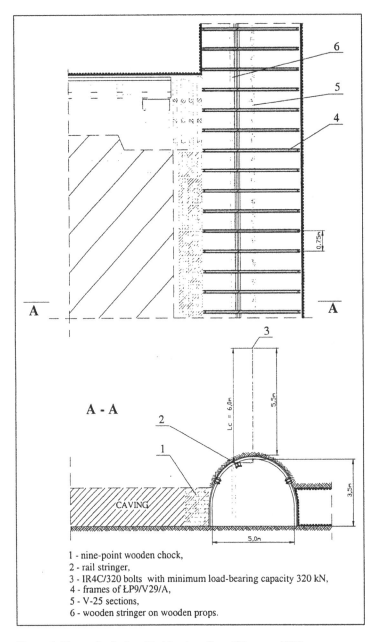

1 - nine-point wooden chock,
2 - rail stringer,
3 - IR4C/320 bolts with minimum load-bearing capacity 320 kN,
4 - frames of ŁP9/V29/A,
5 - V-25 sections,
6 - wooden stringer on wooden props.

Figure 6. The method of roof bolting in gallery j85, seam 407/1

5. OBSERVATIONS OF EXCAVATION STABILITY

Observations of gallery j85 stability were conducted from October 2006 to April 2007. They consisted in measurements of roof rock separation, vertical and horizontal convergence and measurements of frame yield.

At the gallery's j85 life (in terms of distance) amounting to 1170 m 7 bed separation indicators were installed. The heads of bed separation meters were fixed at a distance of about 6.0 m from the roof plane. The measurements consisted in the observation of length changes of a line protruding from the rock mass. The bed separation growth caused by the rise of fractures in the gallery roof appeared in the form of measuring line shortening. Bed separation ranging from 125 to 400 mm was noted. Detailed results were presented in Figure 7.

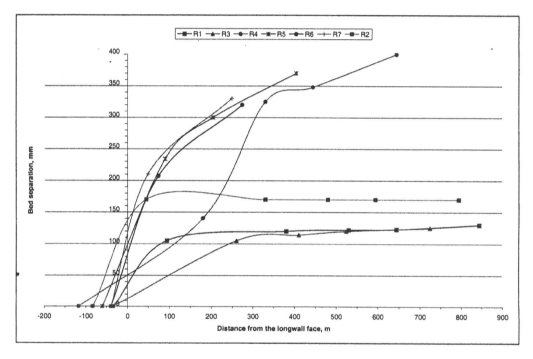

Figure 7. Roof rock separation in gallery j85, seam 407/1

The vertical and horizontal convergence were measured on three consecutive frames located at the distance of about 120 m from the place of longwall j85 run start. Moreover, near every bed separation meter measurements of excavation height and width by help of a measuring tape were carried out (Fig. 8).

Vertical convergence ranging from 600 to 900 mm was noted. Its course in the function of distance from the longwall face was presented in Figure 9.

Horizontal convergence ranged from 200 to 1000 mm. Its course in the function of distance from the longwall face was presented in Figure 10.

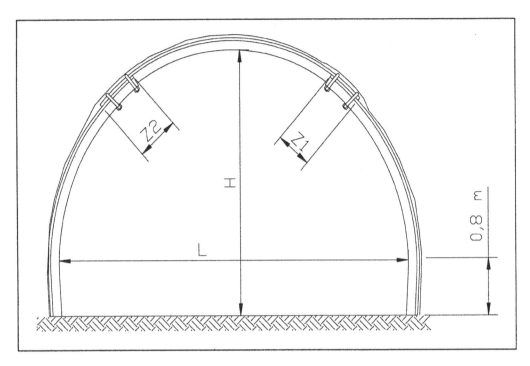

Figure 8. The way of carrying out measurements of convergence and arch yields of frames

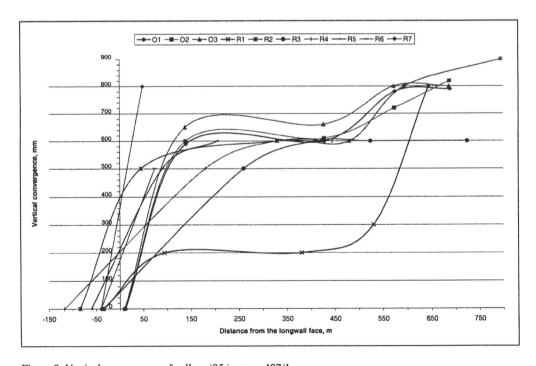

Figure 9. Vertical convergence of gallery j85 in seam 407/1

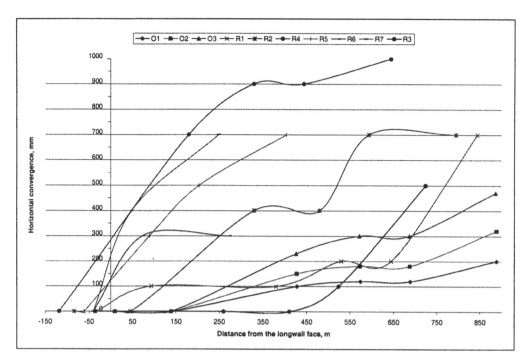

Figure 10. Horizontal convergence of gallery j85 in seam 407/1

The measurements of frame yields were carried out on three consecutive frames located at the distance of about 120 m from the place of longwall j85 run start. The way of measurement carrying out was presented in Figure 8. The first measurements were realised on 4 October 2006. Till 24 April 2007 the following maximum yields were obtained. Frame No. 1: Z2 = 320 mm, frame No. 2: Z2 = 160 mm, frame No. 3: Z2 = 200 mm.

6. ANALYSIS OF OBTAINED RESULTS

When comparing the results of design calculations with the measurement results obtained "in situ", the following statements can be presented.
1. In spite of bed separation measurement results, in the roof of gallery j85 fractures with opening amounting to 400 mm did not originate, but horizontal shift of layers followed.
2. The measurements of vertical convergence were not divided into roof and floor movements. However, basing on underground observations and support arch yield measurements we can state that 70% of excavation vertical convergence constitute floor upheaval and 30% of vertical convergence are roof movements.
3. The predicted in the design by means of the numerical modelling method roof displacement was in fact five times higher.
4. The predicted in the design by means of the numerical modelling method horizontal convergence was in fact ten times higher.

SUMMARY

The application of roof bolting in the gallery j85 in seam 407/1 in order to its maintenance behind the longwall front was entirely successful. The necessity of gallery reconstruction did not occur – all the time its overall dimensions required by regulations were maintained. The above mentioned solution allowed the Mine to achieve considerable savings.

REFERENCES

[1] Biliński A., Kostyk T. 1994: Excavation Support Loads in Longwall Gate Roads. (In Polish), Przegląd Górniczy, No. 6.
[2] Cała M., Flisiak J., Tajduś A. 2001: Mechanism of Bolt Cooperation with the Rock Mass of Differentiated Structure. (In Polish), Library of the Underground Exploitation School 2001, Cracow 2001.
[3] Ciepiela B. 1992: Selected Methods of Maintenance of Excavation's Operational Ability. (In Polish), Wiadomości Górnicze, No. 6.
[4] FLAC v. 4.0 2002: User's Guide. Itasca Consulting Group, Minneapolis.
[5] Jaworski B. 1993: Review of Methods of Safe Maintenance of Longwall Gate Roads. (In Polish), Wiadomości Górnicze, No. 8/9.
[6] Kidybiński A. 1982: Bases of Mine Geotechnology. (In Polish), "Śląsk" Publishing House, Katowice 1982.
[7] Kłeczek Z. 1994: Mining Geomechanics. (In Polish), Silesian Technical Publishing House, Katowice 1994.
[8] Majcherczyk T., Małkowski P. 2003: Impact of Longwall Front on the Size of Fracture Zone Around a Longwall Gate Road. (In Polish), Wiadomości Górnicze, No. 1, Katowice 2003.
[9] Advertising Materials of the Firm IRB Sp. z.o.o.
[10] Matuszewski J., Mąka B., Głuch P. 2006: Maintenance of Longwall Gate Road 23B in the One-Sided Surroundings of Gobs for Longwalls 23 and 24 in Seam 405/1 Mining in Conditions of the "Knurów" Mine. (In Polish), Proceedings of the Seminar on "Problems of Maintenance of Roadway Workings", Silesian Technical University, Gliwice, pp. 174–191.
[11] Nierobisz A., Masny W. 2006: Design of Roof-Bar Arches Fixing and Maintenance of Gallery J85 in Seam 407/1 Behind the Longwall Front. (In Polish), Documentation of research-service work of GIG (non-published), Katowice 2006.
[12] Collective work 1998: Mining Support. Principles of Design and Selection of Roadway Working Support in Hard Coal Mines. (In Polish), Wiadomości Górnicze 1998.
[13] Schissler A.P. 1993: Cyprus Shoshone Tailgate Roof Control – Case Study. Proceedings of the 12th Conference on Ground Control in Mining, WVU (USA), pp. 44–46.
[14] Pilecki Z. et al. 1999: Dynamic Analysis of Impact of Mining Tremor on the Excavation. (In Polish), Puls. Inst. Geophys., Polish Academy of Science, M-22 (310).
[15] Tajduś A., Cała M. 1999: Determination of Parameters of Roadway Working Support on the Basis of Numerical Calculations. (In Polish), AGH, Cracow 1999.
[16] Zimończyk J., Tytko J., Mąka B., Pierchała J., Głuch P. 2006: Solution of Standing Support Reinforcement by Means of a Stringer Fixed to the Roof by Means of String Bolts. (In Polish), Proceedings of the Seminar on "Problems of Maintenance of Roadway Workings", Silesian Technical University, Gliwice, pp. 192–205.

International Mining Forum 2008, Sobczyk & Kicki (eds) © 2008 Taylor & Francis Group, London, ISBN 978-0-415-46126-9

Power Generation Experience from Abandoned Coal Mines Methane

Zaneta Whitworth, S. Cuadrat, Rafał Lewicki
Biogas Technology Limited, Sawtry, Cambridgeshire, United Kingdom

ABSTRACT: Biogas Technology Limited presents two case studies of power generation from abandoned mines methane, using modular generators based on reciprocating spark ignition engines. Results of theoretical modelling and well tests are described together with comparison of the forecast and performance of the units. Experience gained form long term operation of numerous power generation schemes and from leading projects based on the Kyoto protocol allowed to develop business offer in regard to capture and utilization of methane from coal beds, active and abandoned coal mines and air ventilation. Proven in practice testing and operational facilities can be provided, based on unique Biogas business model delivering turnkey solution at no cost to the project partners, from the source testing to realisation of the CERs and energy export.

1. INTRODUCTION

Methane formed during coalification and trapped under pressure in the coal seam is stored within the micro porous coal matrix as a physically adsorbed layer (which usually accounts for 98% of the gas within a coal seam). It will not be released until the coal seam is fractured and pressure is reduced as a result of mining activity or reservoir stimulation such as fracing and de-watering. Once the gas pressure in the coalbed is reduced, the coal seam becomes less capable of retaining methane which then begins to desorb from the surface of the micropores and microfractures into open fractures. The rate of flow is primarily dependent on the diffusion characteristics of the gas and the permeability of the coal seam and adjacent strata.

When a mine is abandoned, the remaining coal may still contain large quantities of methane gas, which, under certain conditions will desorb from the coal into adjacent voids. AMM (Abandoned Mine Methane) schemes aim to extract the methane left in remaining and unminable coal seams.

Methane extraction and utilization in abandoned coalmines can be a cost-effective means of reducing greenhouse gas emissions, whilst contributing to local and regional energy requirements.

In the UK, the exploitation of methane from abandoned coalmines is becoming an established but relatively small industry. Number of successful commercial schemes has been developed to extract and utilize AMM as a source of energy for electricity generation and fuel supply to local industries by dedicated pipelines. Small scale embedded generation projects, based on AMM, Coal Mine Methane (CMM) and Coal Bed Methane (CBM) have been developed thanks to deregulation of energy markets and favourable changes to oil and gas licensing regime. The potential is considerable as there are over 900 closed coal mines in the UK alone with conservatively estimated 300 000 tonnes of methane being emitted to the atmosphere.

Biogas Technology has identified and developed two projects and provided equipment for testing and extracting AMM from coal mines in Llay and Sutton Manor (UK). The experience gained from these projects provides Biogas Technology with opportunities to collaborate internationally in adapting and developing AMM technology around the world.

2. CASE STUDIES

Both sites have been purchased from previous licence owner who decided not to proceed with the project. All the documentation associated with the history of mines and the wells was reviewed and decisions were made to confirm the potential of the wells, drilled into voids of old mineworks as well as investigate options for utilization of the AMM. Tests were commissioned by Biogas Technology and IMC Geophysics Limited. Interpretation of the results and evaluation of the risks indicated viability of the projects.

In order to proceed, quite a few official procedures had to be followed to ensure that the authorities are satisfied with the well development plan, land management, planning permissions, equipment used and end result of the project.

3. PERFORMANCE TEST – AMM RESOURCE AND RESERVE ESTIMATION

The development of an AMM project requires a reliable estimation of the gas volume and production potential of the AMM reservoir.

The key factors are:
– the volume of coal in which the permeability has been enhanced by mining;
– the remaining seam gas content of the coal;
– the volume of gas that can be desorbed from the coal at a given suction pressure;
– groundwater recovery;
– air ingress.

Project life depends on the total available gas volume, the projected annual gas production and the volume of gas made inaccessible by flooding each year. As gas is extracted in time, reservoir pressure will decline necessitating pumping effort to maintain flow. The volume of gas that cannot be recovered due to rising water is obtained by reducing the volume of available methane in proportion to the inflow of water.

The tests involved series of theoretical investigation and modelling based on available information about the coal seams as well as physical testing of the wells including gas flow (controlled venting) and well shut-in exercises, laboratory analyses of gas samples and gas reservoir pressure measurements. The aim was an assessment of potential recoverable gas reserves from abandoned mineworks in order to make commercial decision about developing gas utilization project.

Assumptions were made that rate of gas flow is derived from rate of pressure change data corrected for water pressurization and reduced resource due to mine flooding. Being aware of the restrictions of the modelling, the results had to be interpreted upon extrapolated pressure flow data, uncertain relationship between void volume, gas reserve and flow together with uncertainties of water levels and their change in times.

Biogas Technology designed and supplied hardware for the safe accomplishment of gas flow and recovery tests.

Experience with the well performance tests lead to provision services for other developers interested in CBM ,AMM and CMM projects in respect of manufacturing, supply and commissioning bespoke equipment and instrumentation. Biogas provides also wide range of services in assisting with such well performance tests.

4. SUTTON MANOR FORECAST

Sutton Manor Colliery exploitation started in 1906 and the mineworks were ultimately closed in 1991. Since 1983 the ca 700 m deep Colliery was selling surplus methane gas to nearby chemical company via 5 mile long pipeline, supplying nearly half of fuel required by the ICI Pilkington Sullivan works.

First theoretical investigation in January 2002 was carried out using proprietary software for 3D modelling of minewater recovery and gas reserves.

Well was then drilled into abandoned mine roadway in June 2002. The well was drilled into the main access roadway connected to the pit bottom (roadways around the shaft). Therefore its good connectivity with the bulk of the workings was expected.

Several tests performed consequently gave range of figures that were subject to interpretation and varying certainty levels. The gas at the well contained between 92 and 97% methane and exerted pressure of 2.1 barg.

The estimated void at various stages of investigations varied from some 1.1 to 3.7 million m^3 with lower figures calculated in recent years due to flooding of the mine and probably other uncontrolled events, like collapsing of roadways. Predictions of gas availability at selected extraction and recharge rates differed from 40 weeks to 5 years at various points of tests between June 2002 and December 2005. These were based on the theoretical calculations, like water recharge model and interpretation of physical tests.

Original prediction (2002) indicated that the well will cease gas production due to cutting off connection to remaining gas reserve some in April 2008. It was not possible to estimate the risk of unpredictable events like collapsing of roadway and blockages as mineworks were filling with water. Such events would shorten originally estimated gas production.

Rate of pressure rise in the void calculated from water inflow model was order of magnitude higher than measured in reality. Suspected reasons for this observation were limitations of the model, possibilities of re-adsorption of compressed gas onto the coal within the reserve and gas leakage around the shafts and fault planes. Another risk was isolation of the well by water prior to all the methane being extracted as the well did not access the gas reservoir at highest point.

It is expected that at some point the water overflow the mines roadway. From this moment gas flow will be restricted until being finally cut off when water level will reach the roof if the mine roadway at the base of the well.

Assuming gas extraction rate suitable for 3MW electricity generation, the original expectancy of positive gas pressure was some three years. After this period, forced gas extraction would have to be considered. Tests carried out in December 2005 indicated less than two year gas reserve of gas for 3 MW generation potential.

5. LLAY FORECAST

Construction of the Llay Main Collieries Ltd. started in 1914 and was finally completed in April 1921, initially being just over 800 m deep. The shafts were deepened to 900 m in 1947, making then the colliery the deepest pit in the UK. In 1966 the colliery was closed due to geological problems.

Series of tests and theoretical investigation were carried out between September 2002 and May 2003. The tests had similar limitations, described in case of Sutton Manor. Analysis of gas composition indicated 85% of methane and wellhead pressure of 2.2 barg. Estimates of gas reserves varied for selected scenarios between 3 to 8 years, with the mineworks being completely flooded at some point in the future. Scenarios similar to Sutton Manor were considered.

6. INSTALLATION AND OPERATION

Although the project risk initially seemed to be considerable, a decision has been made to develop both projects with consideration of direct gas supply to the nearby works or on-site power generation and export of the electricity to the grid. Both projects have been developed in partnership with ENERG Natural Power Ltd, a sister company to Biogas Technology.

Gas fuelled engines proven in other applications were selected. The complete sets of modular generators based on low emission Caterpillar 3516 engine are being built in house, at the ENERG Group (sister company of Biogas Technology) workshops in Manchester. The generators are factory pre-commissioned and ready to use after installation on site. Modular design allowed minimizing the risk of the project, developed by stages. The generators are movable between the sites, depending on gas availability and energy production needs.

After securing grid connection supply and satisfying planning permission requirements, the site at Llay (Figure 1) was developed at the end of 2004 at the cost of approximately 1.3 million pounds. Being aware of the limitation of the modelling results, provision was made to move around the generators involved in the project between both sites. Llay started in November 2004 with 2.3 MW with a further 1.15 MW module followed later. The gas plant, connected to the wellhead provided an option for connecting an extraction fan at some stage. The whole installation consists of high and low pressure sections and complied with relevant control and safety requirements. Careful observation of the trends of the gas flow changes and operational regime avoiding cost of forced gas extraction resulted in decision to move one of the generating sets to Sutton Manor at the end of 2006. Production at Llay is planned until at least 2010 with changing the generators to smaller units as and when necessary, down to 300 kW units.

The Sutton Manor site (Figure 2) has been put into operation at the beginning of 2007 also with 2.3 MW capacity that shortly was increased by 1.15 MW generator set brought over from Llay. The site has been developed with budget similar to Llay and the production is planned until the 2012 with the same philosophy of moving the generators.

Figure 1. Llay AMM installation

Figure 2. Sutton Manor AMM installation

Both sites operate unmanned at minimum necessary operation and maintenance costs. All the generation and gas extraction plants are containerised within secure compounds. Telemetry and data recording is linked to the operation centre. Any alarms raised by telemetry system are answered round the clock, depending on the nature of the issue. Periodic preventative maintenance ensures smooth operation of the equipment.

With recently increased energy prices the performance of the projects is more than satisfactory. Calculated risk taken to maximize electricity generation and obtain high return on invested capital proved successful.

Observation of the operational parameters of both projects (Figure 3) gained level of experience in operation of such equipment in rather unpredictable environment. These are compared on regular basis with initial predictions to forecast future of the project. This leads to confidence in further developing of similar schemes along with the Clean Development Mechanism and Joint Implementation, initiated by the Kyoto protocol. Prevention of emissions of methane, potent greenhouse gas, helps towards achieving goals set by the Kyoto protocol and the EU targets. Benefiting the environment, at the same time it helps with energy balance and turns the pollutant into hard cash.

Instrumentation used during the operation of both plants and numerous CDM projects developed by Biogas across the world ensures that any projects developed by Biogas Technology will comply with strict CDM Executive Board regulations and approved methodologies. Parameters measured and logged include electricity exported, methane concentration, gas flow, pressure and temperature, temperature of the flame, should a flare be used to generate emission reductions. Latest and proven type of thermal mass flowmeter is used to obtain the most possibly accurate data to generate Certified Emission Reductions (CERs). The system is operating via telemetry with possibility to control remotely the units. Cost effective equipment maintenance and calibration regime, driven by the UK coordinator ensures data quality and maximisation of the CERs and/or energy generation.

Trained personnel are monitoring performance of the units and reacts locally to any deviations from normal operation of the plant. Following the conditions of the CDM projects, Biogas Technology relies on locally employed staff and creates jobs to run the projects in specific countries. These personnel, trained in the UK, are in continuous liaison with the UK support staff in order to maximise the benefits from the project. Regular reports are subject of ISO 9001-2000 system and internal checks, ensuring infallible approval of independent CDM verifiers.

Indeed, Biogas Technology is one of the world's leading landfill gas project developers.

The company track records speaks for itself:

- the first UK company to develop landfill gas collection and utilisation projects immediately following the implementation of the Kyoto Protocol;
- the first company to be issued verified CERs (Certified Emissions Reductions, better known as carbon credits) from a landfill gas project in Mexico;
- so far the company did not fail any validations and verifications of the projects developed in many countries of the world.

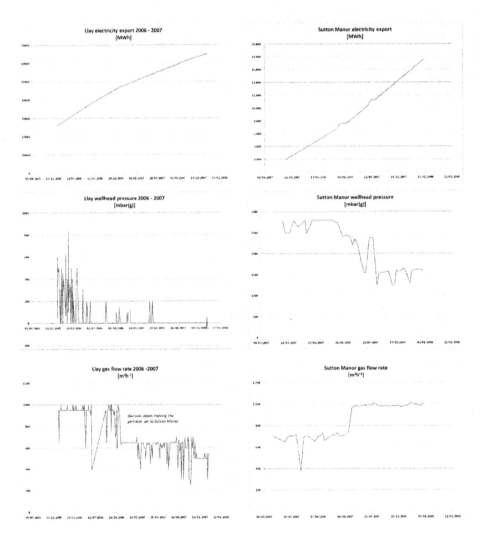

Figure 3. Performance of Llay and Sutton Manor AMM projects

7. JI PROJECT DEVELOPMENT IN COAL MINES

The Kyoto Protocol, which 172 countries have signed so far, was agreed in December 1997 and entered into force in February 2005. It is an international agreement setting targets for industrialized countries to cut their greenhouse gas emissions by 5% compared to 1990 levels, by 2012. The Kyoto Protocol has introduced flexible mechanisms: Clean Development Mechanism (CDM), projects which reduce emissions in countries without a reduction commitment ("non-Annex-I") and Joint Implementation (JI) project mechanism between industrialized and countries in transition, both having a reduction commitment.

CDM & JI are meant to promote projects which lead to reduced emissions of GHG and they allow emissions reductions achieved in one country to be bought by other countries struggling to meet their own targets.

Achieved emissions reductions are popularly called Carbon credits, which are measured in units of certified emission reduction (CER), or Emission Reduction Unit (ERU). Each CER/ERU is equivalent to one ton of carbon dioxide (CO2) reduction.

6 Gases have been identifies as the contributors to Climate change; Carbon Dioxide CO2 – 72% and Methane CH4 – 18% are the major contributors.

Methane (CH4) occurs in a variety of settings; in coal mines it is generated as a by-product of the coal mining process. Methane is a powerful greenhouse gas, 21 times more harmful than CO_2 and its collection and combustion creates "carbon credits". As an illustration, Poland accounts for 3% of the global coal mine methane (CMM) emissions.

Substantial opportunities exist in the coal mining sector to reduce GHG emissions by combusting coal mine methane. This represents a largely untapped energy source that can be used to displace fossil fuels. In many countries, including Poland, large amounts of coal mine methane are currently vented to the atmosphere.

Utilization of CMM represents an opportunity to improve operational safety and environmental performance, as well as potentially providing a clean source of energy for mine operations, and/or for sale to others.

There are a variety of profitable uses for drained CMM and the optimal use at a given location depends on factors such as: the quality of methane, the availability of end-use options, project economics.

High-Quality Gas can be used in natural gas pipelines, for local distribution or vehicle fuel (LNG);

Medium-Quality Gas can be used for power generation, combined heat & power, district heating, coal drying, boiler fuel or other industrial applications; and

Low-Quality Gas & Ventilation Air Methane for oxidation, combustion air and lean burn turbines.

JI project development applies to CMM capture, utilization and destruction at a working coal mine, both new and existing mining activities that involve the use of any of the following extraction activities:
- surface drainage wells to capture CBM associated with mining activities;
- underground boreholes in the mine to capture pre mining CMM;
- ventilation CMM that would normally be vented.

Applicable Methodology is ACM0008: "Consolidated baseline methodology for coal bed methane and coal mine methane capture and use for power (electrical or motive) and heat and/or destruction by flaring".

At present, only methane from active mines is recognized by international targets.

As a consequence, emissions from abandoned mines are not given the same policy priority as methane emissions from active mines.

The Voluntary Emissions Reduction Market has provided the opportunity of earning carbon credits from destruction and utilization of methane in abandoned coal mines.

Such carbon offsets have already been registered and traded on emerging voluntary markets.

Key elements of JI/VER Project development are:
- Technical advisory (e.g. feasibility study);
- Obtaining finance;
- Design, construction, operation and maintenance of the methane collection system, of the flare and/or the utilisation system;
- Project development under the JI or VER scheme;
- Project justification, baseline, studies, public consultation, national approval, registration with the UN or under the VER scheme;
- Trading of CERs or VERs;

As the result of extensive experience, Biogas Technology Ltd. is able to offer:
- Flaring equipment design, construction and sales;
- Turn key provider of Flaring solutions;
- CMM/CBM production testing, providing bespoke systems from wellhead to Flare, including management and technical services;
- Provision of fully funded, owned and operated methane destruction systems as well as Electrical Power Generation from CMM, AMM and VAM;
- JI/CDM project development: One stop – turnkey solution, including
    - Finance,
    - Evaluation (feasibility study),
    - Design,
    - Construction,
    - All necessary equipment,
    - Registration, verification under the chosen scheme,
    - Operation and maintenance for the life of the project,
    - Monitoring,
    - Certification and sale of emissions reductions – CER/ERU or VER,
    - Payment of agreed royalties.

CONCLUSIONS

Undertaking calculated risk of the AMM projects, carefully operated under agreed extraction regime, proved successful.

Operation of the power station demonstrated that AMM is more stable and cleaner fuel source in comparison to landfill gas, resulting in high performance indicators and lower maintenance costs.

Comparison of original forecasts with real data on performance of the projects gives confidence in interpretation of the initial tests and development of further projects.

Experience on development of AMM projects in the UK and CDM projects abroad can be utilised to grow Biogas Technology business in mines methane projects under Kyoto protocol around the world.

Financial independence and business model of Biogas Technology allows delivering turnkey solution at no cost to the project partners, from the source testing to realisation of the CERs and energy export.

Worldwide experience of Biogas Technology brings benefits to the local environment, owner of the gas, local community and ultimately contributes to the global climate change mitigation.

REFERENCES

[1] An Assessment of Potential Recoverable Gas Reserves from Sutton Manor 1 Well. Internal Reports, IMC Geophysics Ltd.
[2] A Further Analysis of Llay 1 Well Reserve. Internal Reports, IMC Geophysics Ltd.
[3] Plant Performance Data at Llay and Sutton Manor. Internal Reports, Biogas Technology Ltd.
[4] http://www.welshcoalmines.co.uk
[5] http://homepage.ntlworld.com/bernard.platt/sutton%20manor%20-%20profile.htm

International Mining Forum 2008, Sobczyk & Kicki (eds) © 2008 Taylor & Francis Group, London, ISBN 978-0-415-46126-9

Author Index

WORLD MINING
CONGRESS
& EXPO 2008

KRAKÓW-KATOWICE
SEPTEMBER 2008

DIAMOND SPONSOR

GOLDEN SPONSOR

MINOVA

KGHM
POLSKA MIEDŹ S.A

Printed and bound by CPI Group (UK) Ltd, Croydon, CR0 4YY

01/11/2024

01782599-0014